RADIO Zeit

RADIO Days

Röhrengeräte
Design-Ikonen
Internetradio

Tube Radios
Design Classics
Internet Radio

KERBER CULTURE

Inhalt

Contents

Vorwort

Preface

Petra Hesse

Das Thema Radio weckt unausweichlich Emotionen und persönliche Erinnerungen. Diese sind meist generationsspezifisch und damit auch immer bezeichnend für den vorherrschenden Zeit- und Musikgeschmack sowie für den jeweils aktuellen Stand der Technik.

Für viele Teenager der 1970er Jahre waren beispielsweise die wöchentlich ausgestrahlten Hitparaden der legendären Jugendsendungen „Radiothek" des WDR oder „Pop Shop" des SWR3 absolute Höhepunkte des persönlichen Musikerlebnisses. Anhänger der Kultsendungen lauschten damals nicht nur gespannt vor dem Radio die nationalen und internationalen Top Ten, sondern versuchten, die heiß begehrten Songs auch zu konservieren, indem sie diese mehr oder weniger professionell mit oder ohne Handmikro auf einem Radio-Kassetten-Rekorder aufnahmen. Die besondere Herausforderung dieser analogen Aufnahmetechnik bestand darin, den Song vollständig ohne Nebengeräusche und Ansage des Moderators aufzuzeichnen, indem man rechtzeitig die Aufnahmetaste drückte, was natürlich nicht immer gelang, dafür aber einen charmanten Charakter hatte. Richtig schlimm war es dann nur, wenn ein Song vom Verkehrsfunk unterbrochen wurde.

Heute ist der Radio-Kassetten-Rekorder – später abgelöst vom Ghettoblaster mit integriertem CD-Laufwerk – genauso wie die bereits 1963 erstmals von Philips auf der IFA präsentierte „Compact Cassette" bereits Medien- und Designgeschichte. Geblieben sind neben den Erinnerungen hierzu die Radiogeräte selbst, die immer häufiger einen festen Platz in Design- und Museumssammlungen finden und teilweise sogar schon als Designklassiker gelten.

Eine der bundesweit größten Radiodesign-Sammlungen befindet sich im Museum für Angewandte Kunst Köln (MAKK). Insbesondere dank der großzügigen Stiftung von Prof. Dr. Richard G. Winkler im Jahr 2005 verfügt das Museum heute über eine Sammlung von 228 Radios, die einen spannenden Überblick über die europäische und amerikanische Geschichte des Radiodesigns geben und zudem – vertreten durch zahlreiche Radiodesign-Ikonen – eindrucksvoll die formale Entwicklung der Radiogehäuse in Hinblick auf Materialität sowie Form- und Farbgebung vermitteln.

The subject of radio inevitably conjures up emotions and personal memories. These are often specific to a generation, representing the corresponding fashions and musical tastes, as well as the then-current technological state-of-the-art.

For example, for many German teenagers in the 1970s, the weekly chart shows of the legendary youth programmes "Radiothek" from WDR or "Pop Shop" from SWR3, were the absolute highlights of their personal music experience. The followers of these cult programmes eagerly sat by their radios, not only to listen to the national and international top ten but also to preserve the coveted songs by recording them on their radio cassette players, more or less professionally, with or without a handheld microphone. The special challenge in this analogue recording technique was to successfully record the whole song without any background noise and without the presenter's announcement. This involved pressing the recording button at just the right moment – an undertaking that sometimes failed but, on the plus side, resulted in recordings of a very special charm. Only when a song was interrupted by traffic announcements, was the recording considered a complete failure.

Today, the radio cassette recorder, which was later replaced by the ghetto blaster with integrated CD player, has long been part of media and design history, as has been Philips' 'compact cassette', launched at the 1963 IFA trade show. What has remained – besides the memories – are the radio devices themselves, which have increasingly found their way into design or museum collections, with some models already being considered design classics.

So zählen zu den unbestrittenen Höhepunkten der Sammlung sicherlich der 1933 von Walter Maria Kersting entworfene „Volksempfänger VE 301", das „Sparton Nocturne" von Walter Dorwin Teague (1935-37), der „Streamliner" von Fada Radio & Electric Co. aus den Jahren 1945/46, der als „Schneewittchensarg" in die Radiogeschichte eingegangene „Phonosuper SK 5" von Hans Gugelot und Dieter Rams (1958), der von Richard Sapper und Marco Zanuso gestaltete „Cubo" von 1964 sowie „Poe" aus dem Jahr 1994 von Philippe Starck (Abb. 1).

Es verwundert deshalb nicht, dass der Ausgangspunkt für die Sonderausstellung „RADIO Zeit. Röhrengeräte, Design-Ikonen, Internetradio" die eigene umfangreiche Sammlung bildet. Verantwortlich für die Idee, das Konzept und die Realisierung der Ausstellung zeichnet Dr. Romana Breuer, im MAKK Kuratorin für Design und Bildende Kunst. Zusammen mit Isabel Brass als wissenschaftliche Volontärin haben beide mit enormen

1 / Poe Radio, Philippe Starck, Thompson Multimedia (FR) für / for Alessi (IT), 1996, Inv.Nr. K 1744, MAKK
© Foto / Photo: Rheinisches Bildarchiv Köln / Cologne, Marion Mennicken.

The Cologne Museum of Applied Arts (MAKK) houses one of the largest radio design collections in Germany. Owing in particular to the generous donation by Prof. Dr. Richard G. Winkler in 2005, the museum boasts a collection of 228 radios, representing a fascinating overview of the history of European and American radio design. Including many icons of radio design, the collection also vividly communicates the development in the design of radio casings, in particular in terms of materials, forms and colour concepts. Among the collection's outstanding highlights are the 1933 "Volksempfänger VE 301" by Walter Maria Kersting, the "Sparton Nocturne" by Walter Dorwin Teague (1935-37), the 1945/46 "Streamliner" from Fada Radio & Electric Co., the 1958 "Phonosuper SK 5" by Hans Gugelot and Dieter Rams – known to radio history by its nickname of 'Snow White's Coffin' –, the 1964 "Cubo" by Richard Sapper and Marco Zanuso and the 1994 "Poe" by Philippe Starck (fig. 1).

It was therefore obvious to use the museum's comprehensive collection as the starting point for the "RADIO Days. Tube Radios, Design Classics, Internet Radio" special exhibition. Dr. Romana Breuer, the MAKK's curator of design and visual arts, developed the exhibition concept and, supported by research assistant Isabel Brass, was responsible for its implementation. With exceptional dedication and commitment and with outstanding expertise, Romana Breuer and Isabel Brass have developed a convincing exhibition concept. Featuring 216 exhibits, the show provides a fascinating tour through the history of radio design. Covering all aspects relevant to design, this also includes, of course, both technological developments and the emotional aspects related to listening to the radio, which are presented in many audio samples and historical recordings due to the archives of WDR, HR and German Broadcast Archiv.

Leistungseinsatz und herausragender Fachkompetenz ein überzeugendes Ausstellungskonzept entwickelt, das anhand von 216 Exponaten einen faszinierenden Rundgang durch die Geschichte des Radiodesigns zeichnet. Dazu gehören natürlich neben allen wesentlichen designrelevanten Gesichtspunkten auch die technischen Entwicklungen und selbstverständlich die emotionalen Bezugspunkte zum Radio, vermittelt durch zahlreiche Hörbeispiele und historische Aufnahmen, die wir vor allem dem Archivmaterial des WDR, des HR und des Deutschen Rundfunk Archivs verdanken.

Begleitend zur Ausstellung entstand eine zweisprachige Publikation, die wesentlich mehr ist als ein klassischer Ausstellungskatalog. Denn wie bereits mehrfach in der Vergangenheit, so wird auch mit „RADIO Zeit" ein Bereich unserer bedeutenden Sammlungen erschlossen, dem Publikum vermittelt und wissenschaftlich dokumentiert. Deshalb beinhaltet der Katalog außer den Beiträgen von Dr. Andreas Baumerich, Isabel Brass, Dr. Romana Breuer, Dr. Dr. h.c. Günter Lattermann, Elina Knorpp und Theresa Nisters zu unterschiedlichen Aspekten des internationalen Radio-Designs ebenso eine vollständige Bestandsdokumentation der gesamten Radio-Sammlung des MAKK. Neben den genannten Autorinnen und Autoren sowie Marion Mennicken vom Rheinischen Bildarchiv für die zahlreichen Objektneuaufnahmen geht mein besonderer Dank an Prof. Dr. Christof Breidenich für die Gestaltung des Kataloglayouts.

Abschließend möchte ich Dr. Romana Breuer und Isabel Brass große Anerkennung und Dank aussprechen. Ebenfalls danken möchte ich dem gesamten MAKK-Team und Werner Nett für die konservatorisch-restauratorische Leitung. Großen Dank schulde ich ebenso der Overstolzengesellschaft – Förderer des Museums für Angewandte Kunst Köln, gegr. 1888 e.V. für die großzügige finanzielle Förderung von „RADIO Zeit" sowie allen Personen und Institutionen, die das Projekt kooperativ mit inhaltlichen Anregungen, mit der Beteiligung am Rahmenprogramm und mit Leihgaben unterstützt und damit ganz individuell ein Puzzleteil zur Realisierung von Ausstellung und Katalog beigetragen haben.

Mein ausdrücklicher Dank geht deshalb an:

In addition to the exhibition, this publication was created, which is much more than just a traditional exhibition catalogue: as with earlier publications, the "RADIO Days" catalogue makes accessible to the public and scientifically documents a particular part of our important collections. Therefore, besides contributions by Dr. Andreas Baumerich, Isabel Brass, Dr. Romana Breuer, Dr. Dr. h.c. Günter Lattermann, Elina Knorpp and Theresa Nisters, which focus on different aspects of international radio design, the catalogue also features a complete documentation of all items included in the MAKK's radio collection. I would like to thank the aforementioned authors, as well as Marion Mennicken from Rheinisches Bildarchiv for the many photographs, and I would also like to extend a special thank you to Prof. Dr. Christof Breidenich for designing the catalogue's layout.

Finally, I would like to especially acknowledge and thank Dr. Romana Breuer and Isabel Brass. I would also like to express my gratitude to the entire MAKK team and to Werner Nett for supervising the conservation and restoration work. Furthermore, I owe particular thanks to the Overstolzen Society – Förderer des Museums für Angewandte Kunst, gegr. 1888 e.V. for its generous financial support of "RADIO Days", as well as to all other individuals and institutions who or which have supported the project, either with ideas and loans or by participating in the supporting events programme. All of these individuals and institutions have contributed an important piece of the puzzle in the implementation of the exhibition and catalogue.

Therefore, I would like to express my sincere thanks to:

Robert Blank, WDR Rundfunkchor, Köln / Cologne

Marion Brass, Viersen

BraunSammlung, Kronberg im Taunus

Manuela Cirillo-Karpf, Viersen

Thierry Didion, Faulx-les-Tombes, Belgien / Belgium

Thomas Edelmann, Hamburg

Werner Ettel, Berlin

Sascha Fuis Fotografie, Köln / Cologne

Thomas Guttandin, Braun Archiv, BraunSammlung,
Kronberg im Taunus

Dr. Herbert Hoven, Köln / Cologne

Dr. Peter Kirchhoff, Fördergesellschaft Rundfunk- und
Tonbandmuseum Köln e.V.

Prof. Dr. Klaus Klemp, Museum Angewandte Kunst, Frankfurt a.M.

Paul Kostial, Köln / Cologne

Karin Lange, Fördergesellschaft RadioMuseum e.V., Köln

Dr. Dr. h.c. Günter Lattermann, Deutsche Gesellschaft für
Kunststoffgeschichte e.V., Bayreuth

Markanto Designklassiker UG, Köln / Cologne

Museum Angewandte Kunst, Frankfurt a.M.

Museum für Kommunikation, Frankfurt a.M.

Museumsdienst Köln / Cologne

Karl Johann Ott, Köln / Cologne

RadioMuseum Köln e.V., Dellbrück

Norbert Rösel, Bonn

Volker Schaeffer, Köln / Cologne

Dr. Marlene Schnelle-Schneyder, Berlin

Stiftung Haus der Geschichte der Bundesrepublik Deutschland,
Bonn

Technoseum, Landesmuseum für Technik und Arbeit,
Mannheim

Sven Vorderstrase, Köln / Cologne

WDR 5

WDR Dokumentation + Archive, Köln / Cologne

WDR STUDIO ZWEI – die Medienwerkstatt

Markus Haßler, WDR Produktionsbetrieb Hörfunk, Köln / Cologne

Petra Witting-Nöthen, WDR Historisches Archiv, Köln / Cologne

Konzerttruhe Komet,
Kuba-Imperial, Wolfen-
büttel (DE), 1957-58,
Stiftung Haus der
Geschichte der Bundes-
republik Deutschland
© Foto / Photo: Ralf
Röttjer

RADIO Zeit – Heldentaten im Design

RADIO Days – Heroic Deeds in Design

Romana Breuer

Das Jahr 1896 markiert einen Meilenstein in der Geschichte der Massenmedien: Im März sendete der russische Physiker Alexander Stepanowitsch Popow (1859-1906) erstmals zwei Worte an eine Empfangsstation. Der Name „Heinrich Hertz" flog 250 Meter durch die Luft. Diese Hommage an Heinrich Hertz (1857-1894) bezieht sich auf dessen Nachweis elektromagnetischer Wellenausbreitung und ihrer Reflexion, der ihm acht Jahre zuvor gelang.[1] Nur einen Monat nach dem geglückten Experiment Popows, im Juni 1896, ließ sich der Italiener Guglielmo Marconi (1874-1937) sein parallel entwickeltes Verfahren zur Übermittlung von Funkimpulsen patentieren. Das Zeitalter der drahtlosen Kommunikation und des Rundfunks hatte begonnen. Die neue Errungenschaft entwickelte sich in den folgenden Jahren in Europa wie in den USA in atemberaubendem Tempo. Ab 1906 ebneten unter anderem der Österreicher Robert von Lieben (1878-1913) und der Deutsche Ferdinand Braun (1850-1918) mit den Entwicklungen der Glühkathodenröhre beziehungsweise des Kristalldetektors den Weg für die private Nutzung der Technologie.

Nachdem 1921 in den USA der öffentliche (Unterhaltungs-) Rundfunk mit 200 vergebenen Sendelizenzen den Startschuss erhalten hatte, wurden auch in Frankreich, Dänemark, England, Russland, Kanada und Argentinien die Hörer zu ersten Rundfunksendungen begrüßt – schließlich am 29.10.1923 auch im Deutschen Reich vom Berliner Vox-Haus aus.[2] Das neue Medium verbreitete sich rasant: Allein die Rundfunkanstalten der Weimarer Republik zählten 1926 schon über eine Million Teilnehmer. Mit wachsender Popularität der Technik entstand aber auch das Bedürfnis in den eigenen vier Wänden – womöglich in der ‚Guten Stube' – alleine oder im Kreise von Familie, Verwandten oder Freunden eine entsprechende Apparatur besitzen und nutzen zu können. Die Faszination für die Allgemeinheit lag nicht nur darin, nun Nachrichten und Informationen direkt zuhause abrufen zu können, sondern auch in der Möglichkeit, Musik zu hören und sogar Geräusche zu empfangen, die vielleicht noch niemals zuvor in der Wohnstube zu Gehör kamen.[3] Jedoch blieb eine Fragestellung in der Frühzeit dieses ersten Massenmediums völlig offen: Wie sieht das Gerät zum Empfang des Rundfunks eigentlich aus? Die rein technischen Komponenten beispielsweise eines Kristalldetektors, die schmucklos auf ein Holzbrett montiert sein konnten, befriedigten zwar das Bedürfnis nach Teilhabe an dem neuen Medium, nicht aber das der Repräsenta-

The year 1896 was a milestone in the history of mass media: in March, Russian physicist Alexander Stepanovich Popov (1859-1906) was the first to wirelessly transmit two words to a radio receiver. The name 'Heinrich Hertz' travelled 250 metres through the ether. This homage to Heinrich Hertz (1857-1894) was due to him proving, eight years earlier, the existence of electromagnetic waves and the fact that they could be reflected.[1] Only one month after Popov's successful experiment, in June 1896, the Italian Guglielmo Marconi (1874-1937) had his method of transmitting radio signals patented, which he had developed at the same time. The era of wireless communication and radio broadcasting had begun. In the following years, in Europe and the USA, the new technology developed at breathtaking speed. From 1906 onwards, among others, the Austrian Robert von Lieben (1878-1913) and the German Ferdinand Braun (1850-1918) paved the way for the private use of the new technology with their developments of the thermionic-cathode tube and the crystal detector respectively.

In 1921, public (entertainment) broadcasting was taking off in the USA with some 200 licenses being granted. France, Denmark, England, Russia, Canada and Argentina would soon also welcome listeners to their first radio programmes. Eventually, on 29 October 1923, the first broadcast was transmitted in the German Reich, from the Vox building in Berlin.[2] The new medium was spreading rapidly: in 1926, the broadcasting companies in the Weimar Republic alone had more than one million listeners. The technology's popularity was constantly growing, and so was the desire to own a radio device and to be able to use it at home, in the living room, either alone or together with family, relatives or friends. The public was not only fascinated by being

tion (Abb. 1). Der Apparat sollte ästhetischen Ansprüchen genügen, um ihn möglichst passend in den Haushalt integrieren zu können.

Die frühen Formen der Gehäuse in Europa und Nordamerika sahen zunächst wie hoch- oder querrechteckige, vielfach eher nüchterne Holz-Kisten aus (vgl. „Radiola 17"[4], 1927/28, Kat.Nr. 2) und hatten daher auch schnell ihre Spitznamen weg: Sie wurden als „Sarkophage" bezeichnet. Das Abhören an diesen Empfangsgeräten erfolgte entweder mittels Kopfhörern oder Lautsprechern. Frühe Beispiele für letztere stellen der BTH Trichterlautsprecher von 1923 (Abb. 2), der amerikanische RCA Lautsprecher (Kat.Nr. 3) zur „Radiola 17" sowie der legendäre Philips Lautsprecher „2007" (Kat.Nr. 4), beide von 1928, dar. Die markante Kreisform mit Spiegel in der Mitte der Lautsprecherversion des niederländischen Herstellers Philips sorgte schnell dafür, dass das Abhörgerät im Volksmund despektierlich „Bratpfanne" genannt wurde.

2 / Trichterlautsprecher / Funnel speaker, BTH, London (GB)
1923 © RadioMuseum Köln e.V.

1 / Detektor Eswe RDN, Sachsenwerk, Niedersedlitz (DE)
1924-25 © RadioMuseum Köln e.V.

able to receive news and information at home, but also by the possibility of listening to music or of receiving sounds that had, perhaps, never been heard before in the living room.[3] However, in the early years of this first mass medium, one question still remained unanswered: what should the devices for receiving radio broadcasts look like? While technical components, such as a crystal detector mounted on a wooden board, met the need for participation in the new medium, these simple devices were not able to fulfil the desire for representation (fig. 1). People wanted devices that also satisfied aesthetic requirements: they wanted the radio to match their interiors.

Gleich mehrere Hersteller widmeten sich in den 1920er Jahren einem aus heutiger Sicht originellen Röhren-Empfangsgerät, das 1924 auf der ersten Funkausstellung in Berlin dem Publikum vorgestellt wurde.[5] Der Apparat besteht aus drei Komponenten,[6] die hintereinandergeschaltet beziehungsweise wie Waggons aneinandergekoppelt sind. Dieser Art der Reihung verdankte die Anlage den Namen „Funk-D-Zug" (Abb. 3).

Aber erst die Zusammenfassung von Empfangsmodul und Lautsprechereinheit führte um 1930 zu weitgehend stilistischer Einheitlichkeit, da nun die Komponenten übereinander montiert wurden. Die ersten Verbindlichkeiten in Bezug auf die Form eines Radios waren getroffen – das Hochformat setzte sich

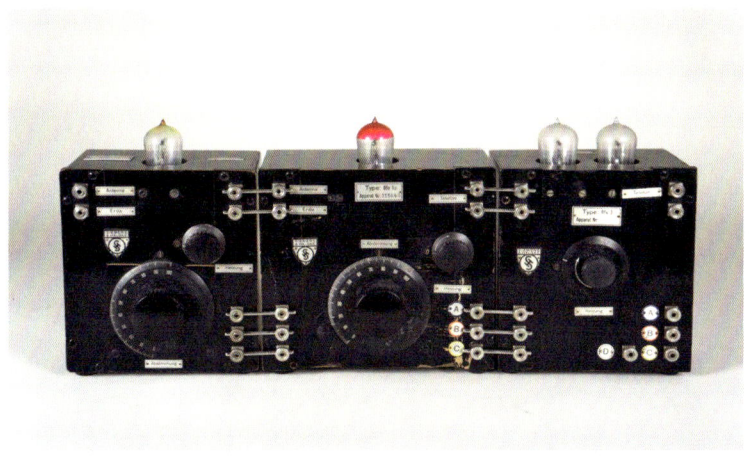

3 / Funk-D-Zug, Siemens & Halske AG, Berlin (DE) 1924
© Museumsstiftung Post und Telekommunikation,
Museum für Kommunikation Frankfurt.

durch. Die frühen Entwerfer orientierten sich dabei häufig an Gebautem, wie auch die Namensgebung für diese Typen verrät. Es gibt die Gruppe der amerikanischen „Tombstones" (Grabsteine), die „Skyscraper" (Wolkenkratzer) und in der gesamten westlichen Welt die „Kathedralen".[7] Zu dieser Gruppe zählt auch der von Walter Maria Kersting[8] entworfene und ab 1933 durch 28 deutsche Hersteller gefertigte Volksempfänger „VE 301" (Kat.Nr. 5 und 6).[9] Obwohl das Gerät äußerlich (wie auch in technischer Hinsicht) eher als schlicht bezeichnet werden kann, lassen sich die Reminiszenzen an Portalbogen und Fensterrosette bei der dezent eingepassten Skala und der übergroßen, kreisrunden Lautsprecheröffnung gut nachvollziehen.

The early casings developed in Europe and North America were vertical or horizontal, often rather plain, rectangular wooden boxes (see "Radiola 17"[4], 1927/28, cat.no. 2) and so quickly acquired the nickname of 'sarcophagus'. Headphones or speakers were used to listen to broadcasts. Early examples of speakers are the 1923 BTH funnel speaker (fig. 2), the 1928 American RCA speaker (cat.no. 3) for the "Radiola 17" and, also from 1928, the legendary "2007" speaker from Philips (cat.no. 4). Due to its distinctive circular shape with a centrally positioned mirror, the speaker by Dutch manufacturer Philips would soon be colloquially known by the rather pejorative name of the 'frying pan'.

In the 1920s, several manufacturers simultaneously developed a valve receiver, which, from today's perspective, is rather original. The device was introduced at the first International Radio Exhibition in Berlin (1924)[5] and consisted of three components[6] connected in series, similar to the coaches of a train. Due to this type of series connection, the device was called the 'Funk-D-Zug' ('radio express train', fig. 3).

Only after receiver and speaker had eventually been combined, around 1930, did a broad stylistic uniformity emerge, as the components were now mounted one above the other. The first rules regarding the form of a radio had been created: the vertical rectangular shape became the dominant design. The early designers of radio casings took inspiration from architecture, which is also indicated in the models' names: there were groups named 'tombstones', 'skyscrapers' and, throughout the western world, 'cathedrals'.[7] Walter Maria Kersting's[8] "VE 301" Volksempfänger (cat.no. 5 and 6), which was produced from 1933 onwards by 28 German manufacturers, also belonged to the 'cathedrals' group.[9] Al-

Die immer noch stetig wachsende Zahl an Rundfunk-Hörerschaft sorgte für einen Boom in der internationalen Radioproduktion. Die Nachfrage konnte mit aufwändig gefertigten Holzgehäusen, die sich auch wegen der Erhitzung der Röhren nur bedingt eigneten, nicht ausreichend befriedigt werden. Abhilfe schufen die seit 1905[10] entwickelten vollsynthetischen Kunststoffe auf Phenolharzbasis, deren Grundstoffe reichhaltig zur Verfügung standen.

Das zunächst unter dem Handelsnamen Bakelit auf dem Markt eingeführte Kunstharz zeichnet sich durch gute Formbarkeit, Hitzebeständigkeit und als Isolator gegen Elektrizität aus.[11] Seit 1927 wurden Radiogehäuse – neben den weiterhin beliebten, hochpreisigen Holzgehäusen – aus Press-Phenolharz produziert, dem die Hersteller mittels Oberflächenveredelung auch eine gewisse „Holzoptik"[12] verliehen. So konnte zum Beispiel das ausgesprochen dekorative Ekco „Consolette RS3" von 1931 (Abb. 4) mit Nussholz- oder Mahagoni-Anmutung erworben werden. Das Radiogerät des britischen Herstellers E.K. Cole Ltd. zählt zudem zu den ersten, die in der Skala die Namen der Sendestationen angaben.

Erik Kirkham Cole entschied sich zu diesem frühen Zeitpunkt, die Produktion der Rundfunkgeräte auf Press-Phenolharz umzustellen, aber dabei – um einem möglichen schlechten Image der Massenware entgegenzuwirken – ein eigenständiges Design zu verfolgen. Für die „Consolette" zeichnet der damalige Chef-Designer bei Ekco J.K. (Jake) White verantwortlich. Darüber hinaus arbeiteten für Ekco auch international bedeutende Entwerfer wie Serge Ivan Chermayeff (1900-1996) und Wells Coates (1895-1958) – beide Architekten gehörten zu den Impulsgebern der Internationalen Moderne in Großbritannien. Das von Coates entworfene Modell „AD 65", mit dem er 1932 einen von der BBC ausgeschriebenen Wettbewerb gewann, besticht durch sein absolut eigenständiges Design, das sich allein schon durch die kreisrunde Form von den üblichen Hoch- und wieder verstärkt auftretenden Querformaten absetzte (Kat.Nr. 7).[13] Beim „AD 65" wie auch dem zwei Jahre späteren Modell „AC 85" (Kat.Nr. 8) wählte Coates einen am Art Déco inspirierten Materialkontrast: Jeweils vor der kreisrunden Lautsprecheröffnung bilden drei senkrechte, verchromte Metallstreben glänzende Zierleisten, die die Wertigkeit des Äußeren betonen. Vergleichbare Akzente setzte

4 / Consolette RS3, Ekco, Southend-on-Sea (GB) 1931
© RadioMuseum Köln e.V.

though the device is rather simple in both looks and technology, references to portal arches and window roses are clearly visible in the subtle design of the dial and the very large circular speaker.

The number of listeners continued to grow, resulting in a boom in international radio production. Demand could not be fully met by the rather costly production of wooden casings, which also, due to the heat given off by the valves, proved to have only limited suitability. This problem was solved by switching to the synthetic plastics made from phenolic resins that were developed from 1905 onwards[10]. The raw materials for these resins were available in abundance.

beispielsweise in den späteren 1930er Jahren die deutsche Löwe-Radio AG mit dem legendären „537 W", das wegen der waagerechten, nach oben links unter der runden Lautsprecheröffnung hochgebogenen Zierleisten den Spitznamen „Schlittschuh" erhielt (Abb. 5). Im Gegensatz jedoch zu den beiden Ekco-Modellen besteht das Gehäuse aus poliertem Holz, erhältlich in einer helleren oder dunkleren Variante.

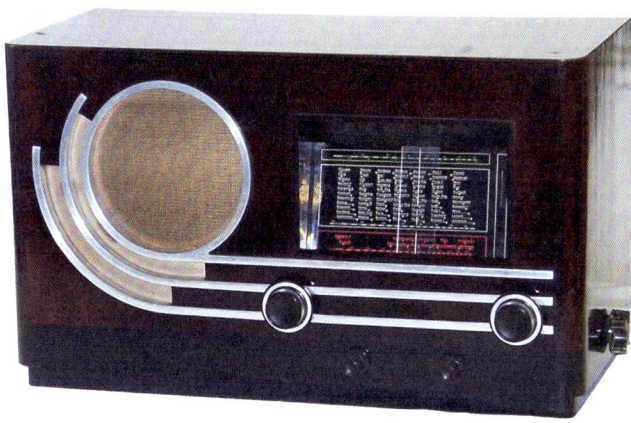

5 / 537 W „Schlittschuh", Löwe-Radio AG, Berlin (DE) 1936-37
© RadioMuseum Köln e.V.

Einen vergleichbaren Ansatz wie die britische Ekco wählte in den 1930er Jahren das US-Unternehmen Sparks-Withington, das seine erfolgreichen Produktlinien seit den beginnenden 1920er Jahren mit dem Kürzel aus beiden Firmenbegründern „Sparton" nannte, indem es auf die exklusiven Entwürfe eines Pioniers des Industriedesigns, Walter Dorwin Teague (1883-1960), setzte. Teague arbeitete jedoch nicht nur mit Kunststoffen, sondern auch mit glänzendem Chrom, Spiegeln, Glas und Emaille. Seine Radios, die zu den Höhepunkten des amerikanischen Art Déco zählen, bestechen sowohl durch die kostbar wirkenden Oberflächen als auch durch die ausgeklügelten Kompositionen. Für das kreisrunde, blauverspiegelte Tischradio „Bluebird" von 1935 (Kat.Nr. 11) durchmaß er den Durchmesser in der Höhe zweifach nach den Gesetzmäßigkeiten des Goldenen Schnitts. Die erste Teilung ergibt den Ansatz des oberen zentralen Chromrings. Die zweite landet exakt im Zentrum der Senderskala.

Initially launched under the trade name of Bakelite, the synthetic resin was characterised by superior moulding properties and heat resistance, while also being an electrical insulator.[11] Alongside the still-popular and costly wooden casings, from 1927 onwards, radio casings were produced from pressed phenolic resin, often with a wood-effect finish.[12] The very decorative 1931 "Consolette RS3" from British manufacturer E.K. Cole Ltd, or Ekco for short, (fig. 4) was, for example, available with a walnut or mahogany finish. It was also the first radio that featured the names of the broadcasting stations on the dial.

Erik Kirkham Cole was one of the first to switch the production of radio devices to pressed phenolic resin and – in order to stave off a potentially negative image for his mass-produced goods – to develop an individual design. Ekco's then chief designer J.K. (Jake) White was responsible for the design of the "Consolette". Other internationally renowned designers, such as the architects Ivan Chermayeff (1900-1996) and Wells Coates, who were among the drivers of international modernism in Great Britain (1895-1958), also worked for Ekco. Coates' "AD 65" radio, which won a 1932 BBC competition, stands out by virtue of its absolutely unique design: its circular shape alone differentiated this radio from the usual vertical or horizontal rectangular boxes (cat.no. 7)[13], the latter having become increasingly popular again. In his design of the "AD 65" and of the "AC 85" (cat.no. 8), developed two years later, Coates had opted for an art déco-inspired contrast in materials: in front of the circular speaker, three vertically arranged chromed metal bars create a reflective décor that underlines the casing's high quality looks. In the late 1930s, the German Löwe-Radio AG company used similar accents in its legendary "537 W". Below the cir-

Mit der seriellen Nutzung von Press-Phenolharzen für die Rundfunkapparate beginnt in Europa, aber besonders auch in den USA, der von dem Kunststoff-Sammler Hans Ulrich Kölsch benannte Matrizen-Stil, der sich unter anderem durch glatte oder geriffelte Oberflächen, abgerundete Ecken und Kanten und kompakte Außenmaße auszeichnet.[14] In welchem Ausmaß der Matrizen-Stil als Kanon für Radios verbindlich war, belegt das Beispiel von Gehäusen, die aus der Design-Schmiede von Ray und Charles Eames stammen. 1946 entwarf die Firma für verschiedene Hersteller Radios mit eben den genannten Merkmalen – aber in geformtem Schichtholz.[15] Die in Europa überaus seltenen und kaum publizierten Radios orientieren sich an der Formensprache der Kunststoff-Gehäuse. Eindrucksvolle Beispiele sind der Vergleich des Zenith „Holiday" von 1945 (Kat.Nr. 31) mit dem Zenith „6 D 030 Z" von 1946 (Kat.Nr. 42) oder des Emerson „678 A" (Abb. 6) mit dem Motorola „55 X 11 A" (Kat.Nr. 40), beide von 1946.

Mit der Weiterentwicklung der noch jungen Kunststoffe – nicht nur Phenolharze sondern beispielsweise auch Harnstoffharze – gelang schließlich die Herstellung eines größeren Farbspektrums. Die American Catalin Corporation hatte die auslaufenden Patente von Bakelit übernommen und entwickelte auf dieser Basis ein Guss-Phenolharz, das in strahlenden Farben und verschiedenen Effekten herstellbar war. Ein vergleichba-

6 / 678 A, Charles und / and Ray Eames, Emerson Radio and Phonograph Corp., New York City, New York (US) 1946,
Markanto Designklassiker UG © Foto / Photo: Rheinisches Bildarchiv Köln / Cologne, Marion Mennicken.

cular speaker, the "537 W" featured horizontal trims, which, on the left hand side, curved upwards, thus giving the radio its nickname of the 'ice skate' (fig. 5). Unlike the two Ekco models, the "537 W"'s casing is made from polished wood and was available in a light or dark finish.

In the 1930s, US company Sparks-Withington – which since the 1920s had named its successful product lines by shortening the founders' two names to 'Sparton' – opted for an approach similar to Ekco's and focused on the exclusive designs by Walter Dorwin Teague (1883-1960). A pioneer of industrial design, Teague not only worked with plastics, but also with reflective chrome, mirror, glass and enamel. Among the most outstanding products of American art déco, Teague's radios are characterised by surfaces that create a luxurious feel and by highly refined compositions. For his 1935 circular "Bluebird" table radio, featuring a blue mirror glass casing (cat.no. 11), he divided the vertical diameter twice according to the laws of the golden section. At the first division, he positioned the upper central chrome ring, while the second division is exactly positioned in the dial's centre.

In Europe, and in particular in the USA, the use of pressed phenolic resin in the production of radio devices marks the beginning of what plastic design collector Hans Ulrich Kölsch has named the 'matrix style'. This style is defined, among other attributes, by smooth or ribbed surfaces, by rounded corners and bevelled edges and by compact dimensions.[14] The radio casings designed by the Ray and Charles Eames studio, for example, show to what degree the matrix style was regarded as a design guideline for radios. In 1946, the studio designed radios for different manufacturers. All of these radi-

rer Kunststoff, der Markenname in den USA lautet Catalin, ist das in Deutschland entwickelte und produzierte Edelkunstharz oder Trolon (beides ebenfalls Handelsnamen). Dieses äußerst flexible Material ließ sich nun in beinahe jedwede Form bringen – ideal für die in den USA einsetzenden Stile des Streamline- wie auch des Machine Age-Designs.

In den folgenden zwei Jahrzehnten boomt die Radioindustrie in Nordamerika, im Nachhinein als „Goldenes Zeitalter des Radios" bezeichnet.[16] Mit dieser Bewertung ist sicher die Wirkung und Ausbreitung des Mediums an sich gemeint, jedoch auch die unglaubliche Qualität und Vielfalt an Gehäusen, die heute gesuchte Sammlerstücke sind. Wunderbare Beispiele für Streamline-Design bis Mitte der 1940er Jahre sind das Belmont Modell „6 D 111", genannt „Rabbit" (Kat.Nr. 38), und das Fada Modell „200"/„1000", genannt „Bullet" (Kat.Nr. 28). Beide Geräte kursieren unabhängig voneinander auch als „Streamliner". Machine Age-Design manifestiert sich in den Gehäusen ebenso wie in den Namen und Bezeichnungen: Das Silvertone „6110" – von Clarence Karstadt 1938 für Sears, Roebuck & Co. entworfen – heißt auch „Rocket" oder „Turbine" (Kat.Nr. 19); Crosley & Division ließ bei der Modellbezeichnung seines „10/135 RR" keine Zweifel über die Inspiration aufkommen. Es trägt den Modellnamen „Studebaker" (Kat.Nr. 65).

Norman Bel Geddes (1893-1958), ebenfalls herausragender Vertreter des Art Déco-Stils und Streamline-Designs, schuf 1940 eines der berühmtesten Radios aus Guss-Phenolharz: das Emerson Modell „400" in den Varianten „Patriot" und „Aristocrat" (Kat.Nr. 25 und 26). Während das „Aristocrat" farblich in verschiedenen, dezenten Marmoreffekten erhältlich war, gab es das „Patriot" ausschließlich in Weiß, Rot und Blau – den Farben des Sternenbanners. Die jeweils nicht für den Korpus gewählte Farbe findet sich auf den Reglerknöpfen, dem Lautsprechergitter und dem Griff des Rundfunkgeräts wieder. Sterne auf den Reglerknöpfen und die waagerechten Lautsprecherschlitze symbolisieren „Stars & Stripes".

Gerade aber am „Patriot" lässt sich aus der heutigen zeitlichen Distanz eine der Schwächen der Guss-Phenolharze ablesen – sie dunkeln nach. Es handelt sich bei diesem Phänomen um eine Luftoxidation, ein chemischer Prozess, der unter ande-

os featured the aforementioned characteristics, but were made from shaped, laminated wood.[15] Hardly known in Europe, these radios are based on the design language of plastic casings. Some very good examples can be found when comparing the 1945 "Holiday" from Zenith (cat.no. 31) with the 1946 "6 D 030 Z" from Zenith (cat.no. 42), or the "678 A" from Emerson (fig. 6) with the "55 X 11 A" from Motorola (cat.no. 40), both from 1946.

By further developing the new plastics technology, using not only phenolic resin but also urea resin, it became possible to create a larger range of colours. The American Catalin Corporation had acquired the expiring patents for Bakelite and, using these as a starting point, developed a cast phenolic resin that could be produced in bright colours and with various finishes. A similar plastic to the one traded under the brand name of Catalin in the US is Edelkunstharz or Trolon (both trade names), developed in Germany. This extremely flexible material could be moulded in almost any form and was thus perfect for the new streamline and machine-age styles emerging in the USA.

Over the following two decades, the North American radio industry experienced a boom, which was later referred to as 'the golden age of radio'.[16] This phrase obviously refers to the effect and spread of the new medium, but also to the incredible quality of, and diversity in, casings, many of which have become sought-after collector's items. Beautiful examples of streamline design right up to the mid-1940s are the "6 D 111" from Belmont, commonly known as the 'Rabbit' (cat.no. 38) and the "1000" model from Fada known as the 'Bullet' (cat. no. 28). Independently of each other, both devices are also called the 'Streamliner'. Machine-age

rem durch die wesentlich geringeren Herstellungstemperaturen (im Vergleich zu den Press-Phenolharzen) erklärbar wird.[17] Besonders im Fall des ehemals strahlend-weißen Gehäuses ist sein heutiger farblicher Zustand verblüffend: Die Oberfläche erscheint bernsteinfarben. Das Blau hat sich durch den verstärkten Gelbanteil des Harzes in Richtung dunkles Grün entwickelt, am unbeschadetsten erscheint das Rot – wenngleich auch hier von einer Nachdunklung ausgegangen werden muss.[18] Erstaunlich mutet in diesem Zusammenhang das immer noch strahlende (wenn auch unterschiedlich ,strahlende') Weiß der Knöpfe, des Griffs und des Gitters an. Die Teile wurden separat von Zulieferern für die Gehäuse gefertigt, es handelt sich vermutlich um Kunststoffteile aus Harnstoffharz.[19]

Während in den USA ab den 1950er Jahren etwas früher als in Europa das Fernsehen als zweites Massenmedium dem Radio seinen Rang streitig machte – und damit den Entwerfern ein neues Aufgabenfeld bescherte –, entwickelten sich die europäischen (und besonders die deutschen) Apparate zunächst in Größe und Material eher zu behäbigen Möbelstücken, die – wer sie sich leisten konnte – als Prestigeobjekt im Wohnzimmer prangten. Die zumeist querformatigen, ,schweren' Geräte aus Holz oder in Holzoptik hatten einige technische wie optische Konstanten: beleuchtete, langestreckte Senderskala inklusive Beschriftung der Senderstationen, große Drehregler links und rechts der Skala, und ,Gebiss-Tasten'[20]. Ein zauberhaftes Exemplar mit eigenständiger äußerer Formgebung stellt die Tonfunk „Violetta Lyra" von 1954 dar (Abb. 7). Trotz der formalen Extravaganz der „Lyra" bleibt die Mehrzahl der Radios dem Kanon des Matrizen-Stils bis weit in die 1950er Jahre verhaftet, wie die abgerundeten Ecken – zumindest an der Front der Holzgehäuse – belegen (z.B. SABA „Meersburg Automatic 9" von 1958-59, Kat. Nr. 100).

Häufig verfügten die Apparate des gehobenen Standards zusätzlich über ein ,Magisches Auge', das zumeist im Bereich des stoffbespannten Lautsprechers angebracht war. Das Magische Auge zeigte an, wie genau ein Gerät auf die Sendefrequenz eingestellt ist. Bei optimalem Empfang leuchtete der gesamte Sektor dieser speziellen Elektronenröhre auf. Den Namen verdankt das Bauteil seiner kreisrunden Form mit mittiger ,Pupille', um die sich die Frequenzfächer schließen. Eine andere Art der Darstellung der Signalstärke ist beispielsweise der Balken, auch ,Magisches Band' genannt.[21]

design manifests in both casings and names or designations: the Silvertone "6110", designed in 1938 by Clarence Karstadt for Sears, Roebuck & Co., is also called the 'Rocket' or 'Turbine' (cat.no. 19). When naming their "10/135 RR" model, Crosley & Division made the inspiration behind the design explicit: the model is called the 'Studebaker' (cat.no. 65).

Another outstanding representative of art déco style and streamline design, in 1940, Norman Bel Geddes (1893-1958) created one of the most famous cast phenolic resin radios: the "400" model from Emerson, in the 'Patriot' and 'Aristocrat' designs (cat.no. 25 and 26). While the Aristocrat was available in different colours with a subtle marble pattern, the 'Patriot' was available only in red, white and blue, the colours of the star-spangled banner. The casing featured one of the three colours while the other two colours were used for the controls, the speaker grille and the handle. The stars on the controls and the horizontal speaker grilles symbolise 'stars & stripes'.

Today, however, one of the weaknesses of cast phenolic resins is showing, in particular in the 'Patriot': it has darkened with age. This phenomenon is due to oxidation by air, a chemical process that results, among other reasons, from the considerably lower production temperatures (in comparison to those used in the production of pressed phenolic resin).[17] The effect is especially strong in the originally bright-white casing: its surface now appears to be amber. Due to the percentage of yellow in the resin, the blue casing has turned to a dark green, while the red resin seems to have best stood the test of time, although we have to assume that it also has become darker.[18] Surprisingly, the whites of the controls, the handle and the grille are still bright,

7 / Violetta W 332 Lyra, Tonfunk GmbH, Karlsruhe (DE) 1954
© Technoseum Mannheim.

Zusätzlich zu den Einzelgeräten boten die Hersteller verstärkt auch Kombinationsmöbel des gehobenen Standards (Radio, Plattenspieler, Tonband, Fernseher oder auch Bar in einem Gehäuse) an, die in ihren Dimensionen zu Sideboards oder Kommoden heranwuchsen. Eine der imposantesten und originellsten Vertreterinnen der Konzerttruhen ist die futuristisch anmutende „Komet", die ab 1957/58 von dem auf Tonmöbel spezialisierten, in Wolfenbüttel ansässigen Unternehmen Kuba-Imperial produziert wurde. Die „Komet" vereint im von vorne gesehen trapezförmigen Unterteil Radio, Plattenspieler und wahlweise auch Tonbandgerät. Das Aufsehen erregende Oberteil, dessen Front ein monumental aufragendes, an eine Haifischflosse erinnerndes, unregelmäßiges Viereck beschreibt, bildet den Rahmen für einen Fernseher. Die verwendeten Hölzer, Kaiserpalme und Ahorn beziehungsweise Wenge und Riegelahorn, stehen in einem reizvollen Helldunkel-Kontrast und beschreiben ein für die 1950er Jahre typisches geometrisches Ornament: einen (asymmetrischen) Doppelkegel mit gegeneinander gerichteten Spitzen.

Die stark eingezogene Taille erinnert an die durch Petticoats hervorgerufene Silhouette der modischen Damenwelt, an Wilhelm Wagenfelds Salz- und Pfefferstreuer-Set „Max und

albeit to varying degrees. These parts were manufactured separately by suppliers, presumably from urea resin.[19]

While in the USA of the 1950s, the new mass medium of television started to challenge the predominance of the radio and thus created new tasks for designers, European (and especially German) radio devices developed into rather immobile pieces of furniture, into prestige objects showcased in the living rooms of those who were able to afford these products. Most of these 'heavy' devices, made from wood, or featuring wood-effect finishes, came as horizontal rectangular boxes and shared particular technical or visual attributes: long backlit dials including the names of broadcasting stations, large rotary controls to the left and right of the dial and 'tooth buttons'.[20] The 1954 "Violetta Lyra" from Tonfunk is a particularly charming example, featuring a unique form (fig. 7). Despite the "Lyra"'s extravagant form, far into the 1950s, the majority of radio devices were still designed according to matrix style principles, which are embodied in the rounded corners that dominated at least the front surfaces of the wooden casings (for example SABA "Meersburg Automatic 9" from 1958-59, cat.no. 100).

High-end products often also had a 'magic eye' that tended to be located in the fabric-covered speaker zone. The magic eye indicated how precisely the radio was tuned into the chosen frequency. When reception was excellent, the whole of this special valve would light up. The component owes its name to its circular shape with a centrally positioned 'pupil', around which the frequency blades would close. Another method of visualising signal strength used a bar shape and was hence called the 'magic bar'.[21]

Moritz" (1952) und steht auch dem amerikanischen Kompakt-radio Emerson Modell „868" (Kat.Nr. 95) von 1956 stilistisch nahe. Ein zeitgenössisches Kuba-Imperial Werbeprospekt siedelte die Konzerttruhe in einem ausgesprochen modischen Interieur mit Cocktailsessel und Nierentischchen an (Abb. 8). Bei der ausgefallenen Chaiselongue, die rechts vorne im Anschnitt zu sehen ist, handelt es sich um den Entwurf des deutschen Architekten Hans Hartl (1899-1980). Das Ruhemöbel zeichnet sich durch eine organische wie asymmetrische Formensprache aus, die sich auch bei den vielfältigen Ausformungen des Nierentischs wiederfindet (Abb. 9).

Mitte der 1950er Jahre bahnten sich in Bezug auf das Äußere eines Radioapparates zwei wesentliche Ereignisse an: Zum einen setzten die Gebrüder Artur und Erwin Braun auf die radikalen, am (historischen) Bauhaus orientierten Ideen der Ulmer Hochschule für Gestaltung[22] – und führten damit den formalisti-

8 / Reklame / Advertisement, „Konzerttruhe Komet", Kuba Imperial, Wolfenbüttel (DE) 1957 © TV-yesterday.

In addition to radio devices, manufacturers increasingly offered high-end combination furniture (radio, turntable, tape recorder, television and even a drinks bar combined in one casing), whose dimensions grew to become sideboards or chests. One of the most impressive and original representatives of phonograms is the futuristic-looking "Komet", produced from 1957/58 onwards by Kuba-Imperial, a company specialising in audio furniture and based in the German town of Wolfenbüttel. The lower part of the "Komet", whose front had a trapezoid shape, housed a radio, a turntable and, optionally, a tape recorder. The upper part with its striking design shaped like a large, irregular flying wedge, rising upwards like a shark fin, forms the frame for the television. The different woods – imperial palm and maple or Wenge and flamed maple – create an attractive light-dark contrast and form a geometric ornament typical of the 1950s: an (asymmetric) double cone whose tips are facing each other. The very narrow waist is reminiscent of a silhouette in women's fashion created by the use of petticoats and of Wilhelm Wagenfeld's "Max und Moritz" salt and pepper shaker set (1952), while also bearing stylistic similarities to the 1956 American "868" compact radio from Emerson (cat.no. 95). A Kuba-Imperial advertising brochure from the time shows the phonogram in very modern interiors featuring a cocktail chair and a kidney-shaped side table (fig. 8). The unconventional chaise longue, partially visible at front right, was designed by German architect Hans Hartl (1899-1980). The piece is defined by an organic and asymmetric design language, also found in the many forms of the kidney-shaped table (fig. 9).

schen Technik-Stil ein. Zum anderen kam in den USA zum Weihnachtsgeschäft 1954 das erste kommerzielle Transistorradio, das „Regency TR-1" auf den Markt.²³ Diese neue und revolutionäre Technik ergänzte beinahe zeitgleich den internationalen Trend

9 / Chaiselongue Soloform 5008, Hans Hartl, Eugen Schmidt GmbH,
Darmstadt (DE) 1953, Inv.Nr. A 2014, MAKK
© Foto / Photo: Manfred Linke.

zum mobilen Kofferradio (Abb. 10). Zwar waren in den USA bereits in den 1920er Jahren erste Versionen von ‚portable radios' in Produktion, aber erst in der Nachkriegszeit wurde das Kofferradio tatsächlich zum Verkaufsschlager. In Deutschland läutete die Firma Grundig 1952 mit dem „Boy Junior", ein Röhrengerät in handlicher, portabler Form, den Trend zur Miniaturisierung und Mobilität ein. In Bezug auf die Technik setzten sich nach und nach die Transistoren durch, gedruckte Platinen, die bis 1969 die Röhren vollständig ablösten. Mit Transistoren ausgestattet konnte jetzt das Radio im Prinzip jedwede, noch so kleine Form einnehmen.

Der durch die Firma Braun eingeleitete Technik-Stil – gemeint ist, dass die Apparate und Komponenten auch im abgeschalteten Zustand nicht mehr hinter Türen, Klappen oder Verblendungen verschwanden – in Kombination mit dem Bestreben nach der „Guten Form"²⁴ verbreitete sich ab den 1960er- und 70er Jahren besonders im High-End-Bereich. Ein schönes Beispiel für die damit einhergehende Kanonisierung einzelner Elemente stellt die Plexiglashaube als Abdeckung für Plattenspieler dar, die seit dem „Phonosuper SK 4" von Braun Standard geworden war.²⁵

In the mid-1950s, two significant developments began to take shape that would strongly affect the design of radio casings: first, the brothers Artur and Erwin Braun focused on the radical ideas of the Ulm School of Design²², based themselves on the (historical) Bauhaus, and thus introduced the techno style. Second, in the USA, the first commercial transistor radio, the "Regency TR-1", was launched in time for the 1954 Christmas season.²³ This new and revolutionary technology coincided almost exactly with the new international trend towards portable radios (fig. 10). Early versions of portable radios had already been produced in the USA in the 1920s, but it was only in the postwar era that the portable radio became a top seller. In Germany, Grundig ushered in the trend towards miniaturisation and mobility with its "Boy Junior" (1952), a small portable valve radio. In terms of technology, by 1969, transistors – printed circuit boards – had gradually replaced valves. Equipped with transistors, the radio could now be given the smallest conceivable form.

With the search for 'good form'²⁴, the techno style introduced by Braun – where devices and components were no longer hidden behind doors, flaps or facings, even when switched off – was becoming increasingly popular in the 1960s and '70s, especially in the high-end segment. A good example of the canonisation of individual elements is the perspex lid for turntables, which, since the launch of the Braun "Phonosuper SK 4", has become the standard.²⁵ As another consequence of the techno style, it had become possible to separate the speakers from the actual device. The techno style met the requirements of an increasingly demanding audience of listeners, who expected optimal sound reproduction (high fidelity) from their combination or

GRUNDIG REISESUPER

GRUNDIG Drucktasten-Boy 54

Neue Röhrentypen mit 25 mA Heizstrom und geringerem Anodenstrom · Batterie- und Netzbetrieb · Röhrenheizung durch gasdichten Stahlsammler, der aus dem Netzteil aufgeladen wird · Schalter für Sparbetrieb · 4 Röhren · 6 Kreise · 3 Wellenbereiche

Eingebaute Ferritstab-Antenne auf allen Bereichen wirksam · Anschlußmöglichkeit für Außenantenne · Übersichtliche Frequenzskala · Feinabstimmung durch Planetentrieb · Kurz-, Mittel- und Langwelle · Röhren: DK 96, DF 96, DAF 96, DL 96 und 2 Trockengleichrichter · Anodenbatterie: 100 Volt · Heizbatterie: Gasdichter Stahlsammler 1,25 Volt oder Monozelle 1,5 V*) · Mittlerer Anodenstrom bei Normalbetrieb 7 mA, bei Sparbetrieb 5 mA · Heizstrom bei Normalbetrieb 125 mA, bei Sparbetrieb 100 mA · Wechselstrom-Anschluß für 110, 125, 220 Volt · Leistungsaufnahme ca. 3,2 Watt · Schwundausgleich auf 2 Röhren wirkend · Permanent-dynamischer Oval-Lautsprecher mit Hochleistungsmagnet 158 x 110 mm · Formschönes Polystyrol-Spritzgußgehäuse mit Metallverzierung und neuartige Zentralbefestigung der Rückwand · Lieferbar in den Farben grün, beige, elfenbein und rot. Abmessungen: 263 x 194 x 91 mm.

DM 186.—

*) Auf Grund unserer laufenden Leistungs-Kontrollen empfehlen wir Ihnen die Batterien der Firmen Baumgarten und Pertrix.
100 V Anodenbatterie:
    Baumgarten Nr. 640, Pertrix Nr. 60.
1,5 V Monozelle:
    Baumgarten Nr. 250, Pertrix Nr. 231.

44

10 / Reklame / Advertisement, „Reisesuper Boy", Grundig-Radiowerke mbH, Fürth (DE) 1953 © Archiv / Archive Radiomuseum Fürth.

component systems. High Fidelity, or HiFi for short, became the new quality standard.[26]

In terms of design, however, only German manufacturer Wega was able to compete with the high benchmark set by the Braun design team around Dieter Rams (born 1932): Wega collaborated with Verner Panton (1926-1998) and Hartmut Esslinger (born 1945), both of whom also created icons of consumer electronics with, for example, the "Stereobar 3300" (Panton, 1970. cat.no. 128) and the "Concept 51 K" (Esslinger, 1975, cat.no. 139).[27]

Consumer electronics manufacturers in other European and non-European countries also relied on the creative potential of designers: among others, Achille and Pier Giacomo Castiglioni (1918-2002 and 1913-1968), Richard Sapper (born 1932), Marco Zanuso (1916-2001) and Mario Bellini (born 1935) worked for the Italian Brionvega company. With their "RR-126-OF-ST" radio-turntable combination, the Castiglioni brothers set a standard for the 1960s and '70s fascination with space travel and robots (cat.no. 120).[28] Two years earlier, Richard Sapper and Marco Zanuso had already created another classic, the "TS 502" foldable radio (cat.no. 117), which is still being produced today, albeit with significantly upgraded technology inside. In 1979, Mario Bellini's "Totem" combination system (designed in 1971, cat. no. 132) won Brionvega the coveted Italian Compasso d'Oro, introduced in 1954.

The pioneering Danish radio manufacturer Bang & Olufsen, which was, and still is, known for premium technological products, also focused on high-quality design by starting their collabora-

Der Technik-Stil bewirkte auch, dass die Lautsprecher wieder das eigentliche Gerät verließen (oder verlassen konnten) und passte zu einer mehr und mehr anspruchsvollen Hörerschaft, die von ihren Kombinations- und Komponentenanlagen hohe Klangwiedergabetreue (engl. High Fidelity) erwarteten. Die High Fidelity, kurz HiFi, wurde neuer Qualitätsstandard.[26]

In Puncto Design konnte aber erst der deutsche Hersteller Wega den hoch angesetzten Maßstäben des Braun Designer-Teams um Dieter Rams (*1932) etwas entgegensetzen: Wega arbeitete mit Verner Panton (1926-98) und Hartmut Esslinger (*1945) zusammen, die beispielsweise mit der „Stereobar 3300" (Panton, 1970, Kat.Nr. 128) und der „Concept 51 K" (Esslinger, 1975, Kat.Nr. 139) ihrerseits Ikonen der Unterhaltungselektronik schufen.[27]

Auch in anderen europäischen und außereuropäischen Ländern setzten die Hersteller von Unterhaltungselektronik auf das Potenzial ihrer Entwerfer: Für das Mailänder Unternehmen Brionvega arbeiteten unter anderem Achille und Pier Giacomo Castiglioni (1918-2002 und 1913-68), Richard Sapper (*1932) und Marco Zanuso (1916-2001) sowie Mario Bellini (*1935). Die Gebrüder Castiglioni setzten mit der Radio-Plattenspieler-Kombination „RR-126-OF-ST" (Kat.Nr. 120) einen Maßstab für die Raumfahrt- und Roboter-Begeisterung der 1960er und 70er Jahre.[28] Bereits zwei Jahre früher gelang Richard Sapper und Marco Zanuso mit dem „TS 502" Klappradio (Kat.Nr. 117) ebenfalls ein Klassiker, der bis heute mit aktualisiertem technischen Innenleben aufgelegt wird. Mario Bellini gewann 1979 mit der Kombinationsanlage „Totem" (Entwurf 1971, Kat.Nr. 132) für Brionvega den seit 1954 verliehenen und begehrten italienischen Compasso d'Oro.

Der dänische Radiopionier Bang & Olufsen, der durch technische Spitzenprodukte bekannt wurde und bis heute ist, setzte seit 1967 mit Jacob Jensen (1926-2015) ebenfalls auf qualitätvolles Design.[29] Er schuf die Basis für die unverwechselbaren B & O Produkte wie beispielsweise das Kofferradio „Beolit 400" (Kat.Nr. 134), das in sechs verschiedenen Farben erhältlich war.[30] Mit dem innovativen Produkt gewann der Designer unter anderem 1971 den iF Design Award. Jensens Nachfolger David Lewis (1939-2011) sollte die Designlinie im Profil noch schärfen. Lewis, der 1995 die hoch angesehene britische Auszeichnung Royal

tion with Jacob Jensen (1926-2015) in 1967.[29] Jensen created the basis for B & O's unique products, such as the "Beolit 400" portable radio (cat.no. 134), which was available in six different colours.[30] This innovative product won Jensen several prizes, including the 1971 iF Design Award. Jensen's successor David Lewis (1939-2011) was contracted to further sharpen the profile of B & O's design language. Lewis, who, in 1995, was awarded the highly regarded British title of Royal Designer for Industry for his B & O products, was also responsible for the 1991 "BeoSound Ouverture" system (cat.no. 153). This audio system not only stands out by virtue of its superior sound, but also by its sensor-controlled interface. The glass doors behind which the individual components are housed are controlled by motion sensors and open, for example, in response to hand gestures.[31]

"Whatever we produce, it has to be the best product in the world", was the credo of the Japanese Sony Corporation.[32] As with the other manufacturers discussed here, the focus was on high-end technology and the matching design, although the company initially did not collaborate with individual (famous) designers. The "TR 1829", designed in 1967 (cat.no. 123), is a good example: the cylindrical shape, which, at the time, was unusual for radios, was meant to be reminiscent of an exclusive table lighter and was, for this reason, also available with a veneer finish. The controls are reduced to a minimum, resulting in a coherent and harmonious look. Launched in 1970, the Sony "ICF 111" (cat.no. 130) was designed for outdoor use and featured several useful and clever details. To protect the control elements from the weather and to prevent breakage when the device was dropped, the control elements are embedded in such a way that they are flush with

Designer for Industry für seine B & O Produkte erhielt, zeichnete 1991 auch verantwortlich für das „BeoSound Ouverture" (Kat. Nr. 153). Das Musiksystem besticht nicht nur durch klangliche Spitzenleistungen, sondern auch durch seine sensorgesteuerte Bedienfront. Die vor den Komponenten angebrachten Glastüren öffnen sich bei Annäherung, also beispielsweise durch eine Handbewegung.[31]

„Was wir auch herstellen – es muss das Beste der Welt sein" lautete das Credo der japanischen Sony Corporation.[32] Wie auch bei den anderen betrachteten Herstellern liegt der Fokus auf anspruchsvoller Technik und ebensolchem Design, zunächst jedoch ohne einzelne (berühmte) Entwerfer im Label zu führen. Ein gutes Beispiel ist das „TR 1829", dessen Entwurf aus dem Jahr 1967 stammt (Kat.Nr. 123). Die für Radioapparate zu diesem Zeitpunkt ungewöhnliche Zylinderform sollte an ein exklusives Tischfeuerzeug erinnern und war aus diesem Grund auch in Furnierholzoptik erhältlich. Alle Bedienelemente sind auf ein Minimum reduziert, so dass sich ein geschlossenes und harmonisches Äußeres ergibt. Das Sony „ICF 111" (Kat.Nr. 130), das 1970 auf den Markt kam, war für den Outdoor-Gebrauch bestimmt und wartete mit einigen zweckmäßigen wie auch intelligenten Detaillösungen auf. Um die Bedienelemente vor Witterungseinflüssen oder der Gefahr des Abbrechens bei einem Sturz zu schützen, sind diese in das Gehäuse so eingelassen, dass sich ein kompaktes Äußeres ergibt. Die knickbare Antenne kann im zusammengeschobenen Zustand auf der gegenüberliegenden Seite arretiert werden, so dass sich daraus direkt auch der Tragegriff bildet.

1974 übernahm Sony Corporation den deutschen Hersteller Wega und konnte so seine Design-Kompetenz besonders mit dem international viel beachteten Designer Hartmut Esslinger auf diesem Sektor noch verstärken. In den 1990er Jahren startete das japanische Unternehmen eine Image-Kampagne auf dem internationalen Markt. Produkte wie „My first Sony" (Kat.Nr. 151), ein Radio-Kassetten-Rekorder mit Mikrofon in fröhlichen, kindgerechten Farben, sollten dazu beitragen, die Kundschaft schon in jungen Jahren an die Marke zu binden. Für die gesamte Produktpalette gab Sony einen Marken-Claim heraus, „It's not a trick. It's a Sony", den die deutsche Agentur KreativRealisten für die gerade nach Deutschland expandierenden Japaner entwickelt hatte.[33] Der schlaue Werbespruch setzte sich in so man-

the casing. When the telescopic aerial was collapsed, it could be folded and locked into the opposite side of the casing to form a handle.

In 1974, the Sony Corporation took over German manufacturer Wega and was thus able to strengthen its design skills in consumer electronics, in particular with the help of internationally acclaimed designer Hartmut Esslinger. In the 1990s, the Japanese company launched an image campaign targeted at the international market. Products like 'My first Sony' (cat.no. 151), a radio cassette recorder with microphone designed in cheerful, child-friendly colours, were aimed at creating customer loyalty at a very early age. Sony introduced a slogan for the entire product range: "It's not a trick. It's a Sony." This slogan had been developed by the German agency KreativRealisten, as Sony had just started to expand into the German market.[33] This clever marketing device stayed in many an ear for a long time. However, the Sony example also demonstrates that, from the mid-1950s onwards, the sector had branched out into two segments: a leisure-focused 'fun' segment (portable radios, miniature radios, toys, promotional products) and a design-focused technology segment (HiFi systems and components). Both segments basically remained constant right up until the 1990s, when they started to face competition by the rapid developments in computer and ICT technologies.[34]

Interestingly, Apple's chief designer Jonathan Ive (born 1967) names, among others, the design guidelines of Braun AG, and explicitly Dieter Rams, as his role models.[35] When comparing, for example, the design of the first 2001 "iPod" with that of the 1956 "Exporter 2" from Braun,[36] there is a striking similarity in the simple circular control element, the

chem Ohr über eine längere Zeit fest. Das Beispiel Sony zeigt aber auch, dass ab Mitte der 1950er Jahren die Branche in zwei Bereiche auseinanderdriftete – den freizeitbetonten ‚Spaß'-Bereich (Kofferradios, Miniaturradios, Spielzeug, Werbeartikel) und den designorientierten Technik-Bereich (HiFi-Anlagen und Komponenten). Beide Ebenen bleiben im Grunde bis in die 1990er Jahre konstant und erhalten dann Konkurrenz durch die sich rasant entwickelnden Möglichkeiten im Bereich der Computer- und Kommunikationstechnologie.[34]

Interessanterweise beruft sich der Kopf des Design-Teams von Apple, Jonathan Ive (*1967), unter anderem auf die Gestaltungsmaximen der Braun AG und ganz konkret auf Dieter Rams als Vorbild.[35] Vergleicht man beispielsweise den ersten iPod von 2001 formal mit dem Braun „Exporter 2" von 1956[36], dann sind das runde, schlichte Bedienfeld, die zurückhaltende Farbigkeit und die abgerundeten Ecken augenfällig. Jedoch auch schon zu dem Tischgerät „Kleinsuper SK 1" und „SK 2" (Kat.Nr. 101) lassen sich solche Parallelen finden. Spätestens an dieser Stelle lässt sich aber für einen möglichen Kanon der Radio-Gehäuse festhalten, dass das Konzept der abgerundeten Ecken – wie es technisch bei den frühen Press- und Guss-Phenolharzen notwendig war – auch bei diesen späteren Design-Ikonen eine Rolle spielt. Die Hersteller Braun in den 1950 und 60er Jahren, Apple seit den 1980er Jahren im Bereich Personal Computer und seit den 2000er in der Multimedia- und Kommunikationsbranche haben sowohl in technischer wie auch in gestalterischer Hinsicht Maßstäbe gesetzt, die sich in der Formensprache der Unterhaltungselektronik stark behauptet haben – wie beispielsweise auch der Blick auf die Produkte des exklusiven Schweizer Sound System Designers Geneva Lab verrät (Kat.Nr. 162).

Ob nun in Zeiten des Internetradios oder der multifunktionalen Smartphones das Radiogerät tatsächlich im Verschwinden begriffen ist? Vielleicht bleibt zur Beantwortung dieser Frage nur ein ‚Jein', um der Ambivalenz der Hersteller und Kundenwünsche gerecht zu werden. Unternehmen wie beispielsweise TechniSat bieten aktuellste DAB+-Technologie sowohl im kühlen, glatten 2000er Design an wie auch im Retro-Stil, der sich an den 1950ern orientiert. Rundfunkanstalten wie Deutschlandradio bieten ihren Hörern eine App in Form einer ‚klassischen' Radiofront an. Es bleibt also abzuwarten, welche nächste Heldentat das Radiodesign erfährt.

subtle colours and the rounded edges. Such parallels, however, can already be found in the "Kleinsuper SK1" and "SK 2" table devices (cat.no. 101). Regarding a possible design canon for radio casings, at this point, we can state that the concept of rounded corners – a necessary feature in the early pressed or cast phenolic resin designs – also plays a role in these later design icons. In both technology and design, Braun (in the 1950s and '60s) and Apple (since the 1980s in personal computing and since the 2000s in multimedia and communication) have set standards that have strongly shaped the design language of consumer electronics, a fact that is also supported when looking at the products from exclusive Swiss audio system designers Geneva Lab (cat.no. 162).

So, are radio devices really disappearing in the era of Internet radio and multifunctional smartphones? In order to do justice to the ambivalent aspects in manufacturers' products and customers' desires, we may be left having to sit on the fence when trying to answer this question. Companies like TechniSat, for example, offer cutting-edge DAB+ technology dressed in both the cool and smooth design of the 2000s and in retro-style casings that take their form from the 1950s. Broadcasters like Deutschlandradio offer their listeners an app designed like the traditional front panel of a radio. We simply have to wait and see what the next milestone in radio design will look like.

1   Vgl. Bremer Rundfunkmuseum, 1992, S. 9.

2   Ebenda, S. 10.

3   Hans Ulrich Kölsch nennt als Beispiele „Meeresbrandung" oder „Löwengebrüll". Kölsch, 1990, S. 5.

4   Trotz des noblen und zurückhaltenden Äußeren handelt es sich bei der „Radiola 17" um ein technisch fortschrittliches Gerät, das als eines der ersten in den USA direkt ans Stromnetz angeschlossen werden konnte, also keiner Batterie mehr bedurfte. Vgl. Schliebe, 2004, S. 35.

5   Die Anlage basierte auf Patenten von Telefunken und wurde beispielsweise von AEG, Siemens & Halske und Stassfurt hergestellt. Freundliche Auskunft von Frank Gnegel, Museum für Kommunikation Frankfurt.

6   Audion-Primär-Empfänger, Niederfrequenzverstärker und Hochfrequenz-verstärker.

7   Ausführlich zu den Architekturbezügen siehe Beitrag Baumerich, S. 27ff.

8   Ausführlich zu Walter Maria Kersting als Entwerfer siehe Beitrag Lattermann, S. 41ff.

9   Das Gerät wurde einheitlich für 75 Reichsmark vertrieben (entspricht heute in etwa 300 Euro) und konnte mit der Stromrechnung abgestottert werden. Später kamen als Produzenten auch sieben österreichische und zwei polnische Hersteller hinzu, die aber ausschließlich die Nachfolgemodelle „VE 301 dyn" (Kat.Nr. 20) und „DKE" (Kat.Nr. 21) produzierten. Diese Entwürfe gehen nicht auf Kersting zurück.

10 Leo Hendrik Baekeland (1863-1944) meldete das von ihm entwickelte Kunstharz 1907 unter dem Namen Bakelit zum Patent an.

11 Vgl. http://www.seilnacht.com/Lexikon/k_phenol.html [8.11.2015].

12 Ulmer, 1997, S. 29.

13 Vgl. Williams, 1997, S. 373-377.

14 Ausführlich zum Matrizen-Stil siehe Beitrag Lattermann, S. 52ff.

15 Neuhart, 2010, S. 330ff.

16 Vgl. Collins, 1997.

17 Freundliche Auskunft und Mitteilung von Günter Lattermann.

18 Ebenda.

19 Ebenda.

20 Der scherzhafte Name bezieht sich auf die weißen bis gelblichen Feststelltasten, die waagerecht nebeneinander angeordnet waren.

21 Magische Augen finden sich bereits an Radioapparaten der 1930er Jahre, wie z.B. der Sparton „1186 Nocturne" (Kat.Nr. 10).

22 Ausführlich zur Designentwicklung bei Braun in den 1950er Jahren siehe Beitrag Knorpp, S. 67ff.

23 Ebenda, S. 60.

24 Ebenda, S. 58.

25 Ebenda, S. 64.

26 Im Produktheft der Firma Grundig von 1972 heißt es dazu: „HiFi (High Fidelity), wörtlich hohe Klangtreue", hebt Stereo auf das Qualitätsniveau der Norm DIN 45 500. [...] GRUNDIG HiFi Geräte übertreffen diese HiFi Norm DIN 45 500 in allen Punkten. Sie garantieren eine Klangwiedergabe in höchster Naturtreue.", GRUNDIG HiFi-Stereo-Programm 72, S. 2.

1   See Bremer Rundfunkmuseum, 1992: 9.

2   Ibid: 10.

3   Hans Ulrich Kölsch uses the examples of ocean waves and a lion's roar. Kölsch 1990: 5.

4   Despite its unobtrusive looks, the "Radiola 17" was a technologically advanced device. It was one of the first radios in the US that could be directly mains-operated and therefore needed no battery. See Schliebe, 2004: 35.

5   The system was based on patents held by Telefunken and produced, among others, by AEG, Siemens & Halske and Stassfurt. Information kindly supplied by Frank Gnegel, Museum für Kommunikation Frankfurt, 10 July 2015.

6   Audion primary receiver, low frequency amplifier and high frequency amplifier.

7   For a detailed discussion of architectural references see essay by Baumerich, pp. 27ff.

8   For more on Walter Maria Kersting's design see essay by Lattermann, pp. 41ff.

9   Later, seven Austrian and two Polish companies also produced Volksempfänger radios, but only the successor models "VE 301 dyn" (cat.no. 20) and "DKE" (cat.no. 21), which were not designed by Kersting.

10 In 1907, Leo Hendrik Baekeland (1863-1944) filed a patent for the phenolic resin that he had developed and named 'Bakelite'.

11 See http://www.seilnacht.com/Lexikon/k_phenol.html [8 Nov. 2015].

12 Ulmer, 1997: 29.

13 See Williams, 1997: 373-377.

14 For more on the matrix style see essay by Lattermann, pp. 52ff.

15 Neuhart, 2010: 330ff.

16 See Collins, 1997.

17 Information kindly supplied by Günter Lattermann.

18 Ibid.

19 Ibid.

20 The jocular name refers to the whiteish-yellow lock keys, which were arranged in a horizontal row.

21 Magic eyes were already used in radios from the 1930s, for example in the Sparton "1186 Nocturne" (cat.no. 10).

22 For more on the design development at Braun during the 1950s see essay by Knorpp, pp. 67ff.

23 Ibid: 60.

24 Ibid: 58.

25 Ibid: 64.

27 Hartmut Esslinger formulierte die Essenz seiner Zusammenarbeit mit WEGA folgendermaßen: „Wir definierten die Designsprache für WEGA mit dem Leitsatz „form follows emotion" und gestalteten das Ganze mit dem Ziel, ein diszipliniertes Gleichgewicht zwischen systemischer Modularität und rundum durchgeformten 360-Grad-Design zu schaffen.", in: Esslinger, 2014, S. 35f.

28 Ausführlich zu Radios im Space-Age-Design siehe Beitrag Brass, hier: S. 76.

29 Sound wird Design, 2011, S.20.

30 Schwarz, Weiß, Rot, Blau, Violett und Curry.

31 „Wenn Sie Ihre Hand ausstrecken um das BeoSound Ouverture zu bedienen, gleiten die Glastüren wie durch Zauber zur Seite und laden Sie zu der Musik ein..." heißt es im Benutzerhandbuch: http://www.bang-olufsen.com/de/customer-service/product-support/sound-systems/beosound-ouverture [16.11.2015].

32 Schmittel, 1975, S. 169.

33 http://www.kreativrealisten.de/kunde/sony/ [16.11.2015].

34 Ausführlich zu der Entwicklung seit den 1980er Jahren siehe Beitrag Nisters, S. 89ff.

35 Heeg/Rams 2010.

36 Vgl. Knorpp, S. 66.

26 In the 1972 Grundig product brochure it states: "HiFi, which means 'high audio fidelity', raises stereo sound to the DIN 45 500 quality standard. [...] GRUNDIG HiFi devices surpass the DIN 45 500 HiFi standard in all points. Our devices guarantee the highest fidelity sound reproduction." GRUNDIG HiFi-Stereo-Programm 72, p. 2.

27 Hartmut Esslinger describes the essence of his collaboration with WEGA thus: "We defined the design language for WEGA with the guideline 'form follows emotion' and designed the products with the objective of creating a disciplined balance between system modularity and an overall, continuously applied 360 degree design." in: Esslinger, 2014: 35f.

28 For more on radios in the space-age design style see essay by Brass: pp. 75.

29 Sound wird Design, 2011: 20.

30 Black, white, red, blue, purple and curry.

31 From the "BeoSound Ouverture" user guide: "As soon as you reach out to touch the operation panel, the glass doors open automatically, revealing the music system concealed behind the screen." http://www.bang-olufsen.com/de/customer-service/product-support/sound-systems/beosound-ouverture [16 Nov. 2015].

32 Schmittel, 1975: 169.

33 http://www.kreativrealisten.de/kunde/sony/ [16 Nov. 2015].

34 For more on the development from the 1980s onwards see essay by Nisters, pp. 89ff.

35 Heeg/Rams 2010.

36 See Knorpp: 66.

Gottesdienst und Luxusleben – Architekturbezüge im Radiodesign

Worship and Luxury – Architectural references in radio design

Andreas Baumerich

Man hört es zuhause, aber es wird im selben Augenblick an einem völlig anderen Ort gesprochen, gesungen oder gespielt.[1] Allein dadurch, dass der Sender der Radiobotschaft unsichtbar bleibt, appelliert das Radio an die Imagination des Zuhörers.[2] Da erscheint es nur passend, wenn auch das Design des Radios oft einen assoziativen Charakter annimmt und es somit durch sein Erscheinungsbild visuell mit den Menschen kommuniziert. Dieser Ansatz wird auch noch dadurch unterstützt, dass die Möglichkeiten zur freien und vielfältigen Gestaltung des Gehäuses, also des ‚Kastens', einen großen Spielraum zulassen.[3] Eine Grundlage dafür lieferte in den 1920er Jahren die Entwicklung des Radios vom technischen Apparat für Bastler zum repräsentativen Kleinmöbel.[4] Die Radios wurden nun oft alleine schon durch ihre Größe zum Bestandteil der Inneneinrichtung,[5] wobei der Einsatz von Architekturelementen eine Möglichkeit war, den neuen Ansprüchen zu genügen. Damit stehen die Radiogehäuse in der jahrhundertealten Tradition, die Kastenobjekte wie Schränke und Truhen durch Bezüge zur hoch geschätzten Baukunst aufzuwerten.

Kathedrale, Kirche, Tempel

Gilt die Baukunst an sich schon als eine anspruchsvolle Referenz, so zeichnet der Bezug zur gotischen Kathedrale[6] als eine der bedeutendsten Leistungen der Architektur das Radio nochmals zusätzlich aus. Vergleichbar wurden auch bei modernen Hochhäusern im Profanbau Elemente der Kathedralarchitektur eingesetzt.[7] Dabei nutzte man im 20. Jahrhundert in der Regel nicht das volle Repertoire mittelalterlicher Architekturgestaltung, um Assoziationen zur Kathedrale hervorzurufen. Das zeitgenössische Publikum kannte wohl die Stile ausreichend, so dass schon stilistische Versatzstücke genügten.

Ähnliche Versatzstücke finden sich dann auch an Radiogeräten in unterschiedlicher Dichte und Deutlichkeit. So steht beim „VE 301 Volksempfänger" (Kat.Nr. 5 und 6), Mende, 1933-35, entworfen von Walter Maria Kersting, die große kreisförmige Lautsprecheröffnung über einem Rundbogen mit der Skala. Trotz der ansonsten sachlichen und funktionsorientierten Gestaltung wecken die beiden Elemente zusammen mit der ‚Hochkantform' des Gehäuses Assoziationen an Fensterrose und Portal mittelalterlicher Kathedralen- bzw. Kirchenfronten wie zum Beispiel der Fassade des Domes von Siena von ca. 1370-80. Der Rundfunk war

You are listening to the radio at home, but, at the same time, what you are hearing is being said, sung or played at a totally different location.[1] The sender of the radio message remains invisible and, by this very fact, the radio appeals to the imagination of the listener.[2] Hence it seems only appropriate that radio design often assumes an associative character by which the device visually communicates with us through its appearance. This approach is additionally supported by the fact that the casing – the box – allows great scope and versatility in design.[3] The basis for this design scope stems from the 1920s, when the radio evolved from being a technical device for hobbyists to becoming a representative piece of furniture.[4] Often, it was simply the size of the radios that turned them into a piece of interior design.[5] The use of architectural elements was a possibility for meeting the corresponding new demands. With this development, radio casings became part of the centuries-old tradition of enhancing box-shaped objects, such as cabinets and chests, by adding visual references from the highly valued art of architecture.

Cathedral, Church, Temple

While using architectural references adds aesthetic value in general, associating the design of radios with one of architecture's greatest achievements – the Gothic cathedral[6] – additionally raises the status of the device. Similarly, elements of cathedral architecture were also used in modern high-rise buildings.[7] In the 20th century, however, it was not necessary to use the entire stylistic repertoire of medieval architectural design to evoke associations with cathedrals. A few stylistic elements would suffice because, presumably, people were sufficiently familiar with the respective styles.

während des Naziregimes ein wirkungsvolles Propagandamittel,[8] wobei das Gerätedesign mit seinen sakralen Assoziationen gut zu den pseudoreligiösen Tendenzen in der NS-Ideologie passte.

Eine dem „VE 301" vergleichbare Reduktion gotischer Elemente zeigt sich häufig im zeitgenössischen Kirchenbau bis ca. 1960. Das Motiv der Fensterrose wird auch hier losgelöst von einer weiterführenden architektonischen Einbindung effektvoll eingesetzt.[9] Einen besonders guten Vergleich zu „VE 301" bie-

1 / Franziskanerkirche St. Marien, Ulrichgasse, Köln / Cologne (DE), Emil Steffann, 1950-1952 © Foto / Photo: Elke Wetzig.

tet die 1950-53 in der Nachkriegszeit entstandene, italienisch anmutende Kölner Franziskanerkirche in der Ulrichgasse von Emil Steffann (Abb. 1).[10]

Weckt das vorhergehende Radio Assoziationen mit einer Kirchenfront, so ermöglicht das „WK Weltklasse" der deutschen Firma Mende, 1933-35 (Abb. 2) quasi einen Blick ins Kircheninnere bzw. in eine gerundete Chorapsis, ähnlich dem Chor der Kölner Minoritenkirche von 1260 (Abb. 3). Die Rundbögen am senkrecht unterteilten Lautsprechergitter übernehmen die Rolle der Chorfenster und die Kreissegmente der Skala darunter stehen für den Fußboden. Hier noch die Verbindung der Begriffe Chorapsis und Sängerchor mitzudenken, macht natürlich auch des-

In varying degrees of obviousness and density, similar elements can also be found in radio design. For example, in the "VE 301 Volksempfänger" (cat.no. 5 and 6) from Mende, 1933-35, designed by Walter Maria Kersting, the large circular speaker is positioned above an arch that houses the dial display. Despite the radio's overall simple and functional design, the circular speaker and arch, together with the casing's vertical-rectangular form, are reminiscent of the window roses and portals of medieval cathedrals or churches, for example of the façade of the Cathedral of Siena (ca. 1370-80). In Germany, during the Nazi regime, radio was an efficient means of propaganda[8] and radio design, with its religious references, perfectly matched the pseudo-religious tendencies in Nazi ideology.

2 / WK Weltklasse, Mende & Co. GmbH, Dresden (DE) 1933-35., Paul Kostial © Foto / Photo: Rheinisches Bildarchiv Köln / Cologne, Marion Mennicken.

wegen Sinn, weil das Radio eine Quelle des – meist allerdings nicht sakralen – Gesangs ist. Rein formal hat wahrscheinlich schon die ‚Hochkantform' bei vielen Modellen die Verbindung mit einer Kirchenfront nahe gelegt.[11] Sicher war sie auch passend, weil dem Radio etwas Spirituelles, Magisches bzw. Übersinnliches anhaftete und die körperlosen Radiostimmen sogar als singende Engel in der Luft beschrieben wurden.[12] Das sakral inszenierte Radio wurde dann oft auch noch an einer zentralen Stelle im Wohnraum aufgestellt und zu den Sendungen versam-

3 / Chor / Choir, Minoritenkirche, Köln / Cologne, 1260.

melten sich die Bewohner wie die Gemeinde um die Kanzel oder den Altar. Kein Wunder also, dass die Kirchen die Konkurrenz des neuen Mediums beklagten.[13]

Eine nicht christliche, aber dennoch weihevoll-sakrale Assoziation ermöglicht das grau-rote Radio Modell „5A-2" der US-Firma Garod von 1946 (Kat.Nr. 36). Es erinnert in reduktionistischer Weise an eine Tempelfassade. Das seit der Renaissance als Hoheitsmoment genutzte Baumotiv wurde bei Kirchen, aber auch zur Bedeutungshebung profaner Gebäude eingesetzt. Gerade in den USA hat der Tempelportikus als Teil von öffentlichen und privaten Bauten seit Gründung des Landes eine besondere Tradition.[14] In diesem US-amerikanischen Zusammenhang fühlt

Similar to the above-described design of the "VE 301", the use of individual Gothic elements can often be found in church architecture from the 1930s to the 1960s. Here too, the window rose is lifted from its original architectural context and used to great effect.[9] Designed by Emil Steffann and built during the postwar period, the Italian-looking Franciscan church on Cologne's Ulrichgasse (fig. 1) is a very good comparison to the "VE 301" radio.[10] While the "VE 301" is reminiscent of a church façade, the "WK Weltklasse" from German company Mende (1933-35, fig. 2) provides, if you like, a view into a church or rather into a curved choir, similar to the choir in Cologne's Minorite Church, which dates from 1260 (fig. 3). The two arched shapes in the radio's vertically divided speaker grid look like choir windows, while the circular segments of the dial below represent the floor. Associating terms like 'choir' with this design makes sense because the radio can also be a source of singing, albeit mostly of the non-religious variety. The vertical-rectangular form used in many radio designs already suggests a reference to church façades.[11] This reference also seems fitting because the radio used to have a spiritual, magical or supernatural aura and the disembodied radio voices were even described as "singing angels in the air".[12] With its ecclesiastical appearance, the radio was often positioned at a central location in the living room, with people gathering around it to listen to broadcasts like worshippers gathered around the pulpit or altar. Not surprisingly, the churches were unhappy about having to compete with the new medium.[13]

While not relating to Christian traditions, the 1946 grey-red "5A-2" radio from US company Garod (cat.no. 36) nonetheless evokes solemn-sacred associations. In a reductionist way, it is reminis-

man sich bei der Gestaltung von Modell „5A-2" an die Front des Lincoln-Memorials von 1915-22 in Washington, D.C. von Henry Bacon (Abb. 4), erinnert, dessen Säulenfront einen Architrav aber keinen Giebel besitzt.

Durch die an Scheinwerfer erinnernden beidseitigen Reglerknöpfe ähnelt das „5A-2" aber auch der prominenten ‚Tempelfassade' beim Kühlergrill des britischen Automobilklassikers Rolls-Royce.[15] Auch bei ihm verbinden sich Repräsentation und Technik. In beiden Fällen steht die Fassade für eine aus der Tradition erwachsenen Bedeutung und Noblesse.

4 / Lincoln-Memorial, Washington, District of Columbia (US), Henry Bacon, 1915-22 © Foto / Photo: Martin Falbisoner.

Hochhaus, Monument, Grabstein

Die beim damaligen Radiodesign verbreitete ‚Hochkantform' erlaubte es ebenso, an den zeitgenössischen Hochhausbau anzuknüpfen. In einer von Technik begeisterten Zeit stellten Hochhäuser eine bewunderte technische Höchstleistung dar, die in den 1920er und 1930er Jahren neue Maßstäbe setzte, wie sie teilweise erst Jahrzehnte später übertroffen wurden.[16] Damit ließ sich in den Vereinigten Staaten wohl auch ein Radiogerät mit Anklängen an Hochhausarchitektur als ein patriotischer Verweis auf nationale Leistungen lesen.[17] Die Gestaltung der Hochhäuser wurde dementsprechend auch von den Designern als „kraftvolles Bild"[18] erkannt und genutzt.[19] Ein besonders eindrucksvol-

cent of a temple façade. A symbol of sovereignty since the renaissance, the temple motif was used in churches, but also to heighten the importance of secular buildings. Especially in the USA, the temple portico has featured in public and residential buildings since the nation's founding.[14] In the US context, the design of the "5A-2" radio is reminiscent of the front of Henry Bacon's Lincoln Memorial (1915-22) in Washington D.C. (fig. 4), whose columnar façade sports, in place of a pediment, a massive trabeated frieze. However, with its two control dials being reminiscent of headlights, the "5A-2" is also similar to the prominent 'temple façade' featured in the grille of the British car classic, the Rolls-Royce.[15] Here too, representation and technology are combined. In both the "5A-2" and the Rolls Royce, the face or façade symbolises importance and noblesse derived from tradition.

Skyscraper, Monument, Tombstone

The vertical-rectangular form popular in the radio design of the 1920s and '30s was also a reference to the then-contemporary construction of high-rise buildings. In a time characterised by fascination with new technologies, high-rise buildings were admired as stand-out technological achievements which, in the 1920s and 30s, set totally new standards, some of which would only be surpassed decades later.[16] Hence, in the United States, radio design alluding to high-rise architecture could also be interpreted as a patriotic reference to national achievement.[17] Consequently, designers recognised and used the "powerful image"[18] of high-rise architecture.[19] A particularly impressive example is the "66" radio from Air-King Products Company, designed by Harold L. van Doren and John Gordon Rideout in 1935/36 (cat.no. 9) The horizontal ribs at the base can be interpreted as a reduced form of

les Beispiel hierfür ist das Radiomodell mit der Nummer „66" der Air-King Products Company, das von Harold L. van Doren und John Gordon Rideout 1935/36 entworfen wurde (Kat.Nr. 9). Horizontale Rippen im Sockelbereich lassen sich als Reduktionsform einer Rustikaquaderung lesen.[20] In der Mitte werden sie von einem Band mit senkrechter Riffelung unterbrochen. Dieser Streifen geht über die gesamte Höhe der Front und setzt sich in die ‚Dachfläche' hinein fort, wo sich die schlitzförmigen Öffnungen des Lautsprechers befinden.[21] Da ansonsten auf Ornament verzichtet wurde, kommt der Riffelung als Gestaltungsmittel eine wichtige Rolle zu. Sie stellt ein typisches Motiv des Art Décos und der Streamline-Moderne dar.[22] Bei „66" verweisen die Schlitze an der Oberseite zusätzlich auf den praktisch-technischen Bereich. Typisch für das Art Déco verbinden sich hier Tradition, Modernität und Technikbegeisterung.

Dem geriffelten Streifen ist zunächst eine kreisförmige Skala aufgesetzt. Durch ihren oben und unten gestuften Rahmen erinnert sie an einen vorgesetzten Stufengiebel, wodurch sie mit dem Stufenabschluss des Kastens korrespondiert. Es folgt nach oben ein rechteckiges Fenster mit einer Darstellung der beiden Hemisphären. Dieses Weltläufigkeit suggerierende Motiv ist bei anderen Varianten des Modells durch eine Uhr ersetzt.[23] Einen vergleichbar pointierten Einsatz einer Uhr zeigen auch einige Hochhäuser, die – wie das Gerät – dadurch quasi zum Uhrenturm werden.[24]

5 / Los Angeles Times Building, Los Angeles, Kalifornien / California (US), Gordon B. Kaufman, 1931-35 © Foto / Photo: Minnaert.

rustication.[20] In the centre, the horizontal ribs are interrupted by a vertically ribbed band along the entire length of the front surface that continues to the 'roof surface' where the speaker slits are located.[21] As there is no other ornamentation, the ribs are an important design feature, typical of art déco and modernist 'streamline' design.[22] In the "66" model, the slits at the top of the case are also references to the practical-technological area. Typical for art déco, the design of the "66" radio unites tradition, modernity and fascination with technology.

In the lower part of the radio's front, a circular dial is mounted on top of the ribbed band. With its stepped bezel, the dial is reminiscent of a protruding stepped gable and mirrors the stepped shape of the casing. Above the dial is a rectangular window showing an illustration of the two hemispheres of the globe that alludes to cosmopolitanism. This motif is replaced by a clock in other designs of the "66" model.[23] Some high-rise buildings also feature similarly prominently positioned clocks, thus turning the buildings – like the radio – into clocktowers.[24] Stepped shapes or outlines, as used in the design of the "66" radio, were characteristic features of art déco buildings.[25] The basic structure and detailing of the Los Angeles Times building (1931-35) by Gordon B. Kaufman are similar to the "66" radio (fig. 5). There were also quite practical reasons for using a stepped design[26], which, in the US, can take its form from Mexican or Egyptian step pyramids or from Babylonian ziggurats, depending on the individual design.[27] According to Johann N. Schmidt, referencing historical monuments was an expression of the New World's desire "to morphologically connect to heroic times."[28] The radio created this heroic connection for the living room.

Für Art Déco-Bauten waren Stufungen wie bei dem Radio-modell „66" charakteristisch.[25] Ihm ähnelt in seiner Grundstruktur und seinen Details das Los Angeles Times Building von Gordon B. Kaufman aus den Jahren 1931-35 (Abb. 5). Die Stufung war durchaus auch praktisch begründet,[26] kann aber in den USA je nach Gestaltung mexikanische bzw. ägyptische Stufenpyramiden oder babylonische Zikkurats als Vorbild haben.[27] Die Anknüp-fung an historische Monumente verdeutlichte nach Johann N. Schmidt, dass die Neue Welt „auch morphologisch den Anschluss an heroische Zeiten"[28] suchte. Das Radiogerät stellte diesen ‚heroischen Anschluss' für das Wohnzimmer her.

Das Radiomodell „66" wird zwar als „Skyscraper", Wolken-kratzer, bezeichnet, aber seine gedrungenen Proportionen sind eher die eines hohen Hauses. Eine kleinere Standfläche hätte wohl eine größere Instabilität erzeugt. Dennoch geht bei „66" der Gedanke in Richtung Hochhausarchitektur, da das Hochhaus mit Abstufungen der wohl auffallendste und bekannteste Bautyp des Art Décos war. Und beispielsweise aufgesetzt auf ein Kasten-möbel mutete es in der häuslichen Umgebung sicher noch mehr wie ein Miniaturhochhaus und damit Monument an. Die akus-tische Aufmerksamkeit hatte es im angeschalteten Zustand so-wieso; die visuelle erhielt es durch seine Position und auffallen-de Gestaltung. Das Hochhaus und auch weniger hohe Gebäude des Art Décos zeichnen eine Tendenz zum Monumentalen aus, die sich in schweren Formen und Anleihen an Sakral- oder Grab-bauten wie Tempeln,[29] Kirchen [30] oder Pyramiden[31] äußert. Die ‚Hochkantform' und die optisch ‚schwere' Gestaltung moch-ten bei Radiogeräten aber auch Assoziationen zu Grabsteinen quasi als Kleinstmonumenten wecken.[32] Mit dem Monument ist ein zeichenhafter Charakter verbunden, der auch den bedeuten-den Hochhäusern anhaftet. Ihre Dynamik und Fortschrittlichkeit passen zur technischen Seite des Radios. Ihr Ausdruck von wirt-schaftlicher Potenz und Geltungsbedürfnis[33] findet seinen Spie-gel im Kostenaufwand für ein Radiogerät sowie in dessen archi-tektonisch aufgewerteter Gestaltung.

Dampfer, Haus, Fenster

Die Lautsprecheröffnungen der Radiomodelle „A5" der Firma Halson von 1938, entworfen von J. Samson Spencer (Kat.Nr. 17) und „Transitone TP 10" der US-Firma Philco von 1939 (Kat.Nr. 18) haben Ähnlichkeit mit um eine Hausecke gezoge-nen Fensterbändern. Ihre Skalen erinnern an Bullaugenfenster.

Although the "66" radio is referred to as a 'skyscraper', its squat proportions are rather those of a tall house. A smaller footprint would likely have re-sulted in greater instability. Nevertheless, the "66" is reminiscent of high-rise architecture, as stepped high-rise structures were the most striking and well-known designs of the art déco style. When, for example, placed on top of a box-shaped piece of furniture at the home, the "66" radio would have looked even more like a miniature high-rise build-ing and, indeed, a monument. When switched on, it would have been the focal point of acoustic at-tention; with its striking design and prominent placement, it was also guaranteed to be the cen-tre of visual attention. High-rise buildings, as well as other, lower structures from the art déco period, are characterised by a tendency towards the monu-mental, expressed in forms that exude gravitas and in references to sacred or funerary structures such as temples[29], churches[30] or pyramids[31]. The verti-cal-rectangular form and the visual 'gravitas' may also have evoked images of tombstones, which are akin to miniature monuments.[32] Monuments have symbolic character, as do signature high-rise build-ings. Their dynamic qualities and expression of technological advancement correspond to the tech-nological status of the radio. The economic prow-ess symbolised by high-rise buildings and their at-tention-demanding[33] presence are mirrored in the prices that had to be paid for radios and in their de-sign enhanced by architectural references.

Steamliner, House, Window

The speaker slits in both the 1938 "A5" radio from Halson, designed by J. Samson Spencer (cat.no. 17) and in the 1939 "Transitone TP 10" from US company Philco (cat.no. 18) are reminiscent of a window strip bending around the edge of a build-

6 / Hotel Crescent, Miami Beach, Florida (US),
Henry Hohauser, 1932 © Foto / Photo: Jean-François Loiseau.

Die Motive der Fensterbänder[34] und Bullaugenfenster[35] gehören zum Repertoire moderner Architektur verschiedener Richtungen und können separat oder in Kombination erscheinen. So zeigt beispielsweise das 1932 fertiggestellte Hotel Crescent in Miami Beach von Henry Hohauser (Abb. 6) eine deutliche Verwandtschaft mit der Gestaltung des „A5".

Beide Architekturmotive verweisen auf Ozeandampfer.[36] Diese boten einen vielschichtigen Assoziationsrahmen von Technik, Luxus und Eleganz, der sowohl traditionalistische als auch modernistische Architektur beeinflusste.[37] Der mit den Dampfern verbundene Traum von der ‚großen weiten Welt' war bei den beiden vorhergehenden Radios durch ihr relativ preiswertes Kunststoffmaterial auch für den kleineren Geldbeutel zu befriedigen. Deutlich näher zur Luxuswelt stehen dagegen die von Walter Dorwin Teague entworfenen Modelle Nr. „557" und „558", genannt „Schlitten", der US-amerikanischen Sparton Sparks-Withington Company von 1937-38 (Kat.Nr. 13 und 14). Zunächst bestechen beide Radios durch ihre Spiegelumhüllung, wie sie in den 1930ern gerade in den Farben Apricot und Blau be-

ing. The dial displays of both models are similar to art déco's beloved porthole-shaped windows. The motifs of window strip[34] and porthole window[35] belong to the repertoire of different styles of modern architecture, where these two features are used either individually or in combination. For example, the Crescent Hotel in Miami Beach, designed by Henry Hohauser and completed in 1932 (fig. 6) bears a striking similarity to the "A5" radio.

The architectural motifs of window strip and porthole window are a reference to steamliners.[36] These ships provided a multi-layered associational framework, alluding to technology, luxury and elegance, which influenced both traditional and modernist architecture.[37] The dream of 'the big wide world' symbolised by these steamships could be bought at a comparably small price by acquiring one of the two above-mentioned radios, whose cases were made of inexpensive plastic. Compared to these radios, the 1937 "557" and 1938 "558" 'sled' radios designed by Walter Dorwin Teague and produced by the US company Sparton Sparks-Withington were much closer associated with the world of luxury (cat.no. 13 and 14). Both radios feature a striking glass casing, which was popular in the 1930s, in particular in the colours apricot and blue.[38] The casing corresponds, albeit on a smaller scale, to the then-glamorous style of furniture with high-gloss surfaces.[39] One could also think of the large glass façades in Bauhaus-modernist buildings or the colourful and glossy ceramic cladding of art déco structures.[40] With regard to the "457X" design of the sled model, Attwood points to Le Corbusier and other European modernists as being role models for Dorwin Teague.[41] On the one hand, the horizontal lines of the radios' protruding, vertically arranged glass fins are reminiscent of a steamliner's

sonders populär waren.[38] Dies passt zur zeitgenössischen Mode der Spiegelmöbel, an deren Glamour die Radios im Kleinen teilhaben.[39] Man könnte aber auch an große spiegelnde Fensterflächen von Gebäuden der Bauhaus-Moderne oder farbig-glänzende Keramikverkleidungen von Art Déco-Bauten denken.[40] So verweist Attwood im Zusammenhang mit der Version „457X" des ‚Schlittens' auf die Vorbildrolle Le Corbusiers und anderer europäischer Modernisten für Teague.[41]

7 / Seagram Building, New York City, New York (US), Ludwig Mies van der Rohe, 1958 © Foto / Photo: Steve Cadman.

stacked decks. On the other hand, the lines could also be a reference to the architectural motif of 'eyebrows' above windows, as can be found in designs by Frank Lloyd Wright, in art déco buildings or in 'streamline' modernist structures.[42] The motif of rows of portholes represented by the control dials below the 'deck' can also be found in architecture.[43]

The "6 D 030 Z" radio from Evans Plywood Company for Zenith (cat.no. 42), on the other hand, features the prominent use of traditional architectural elements. The radio was designed in 1946 by Charles Eames and his wife, the sculptor Ray Eames. The classic architectural element of the semi-circular arch is used upside-down while the wedge-shaped 'capstone' divides the arch's surface. The wedge becomes an arrow pointing towards the control dial. A similar use of architectural elements can often be found in 20th century architecture. Disassociated from their original function, these elements are given a symbolic character that, while alluding to the historical context, is used in a playful way.[44]

Designed in 1949 by Raymond Loewy, the "602 A" model from US company Emerson Radio and Phonograph Corporation uses a typical modernist window design (cat.no. 60): with its bulky form and a speaker grille featuring deep openings, the radio looks like a scaled-down model of a building with a grid façade. These types of buildings define the urban landscapes of large modern cities, like, for example, Ludwig Mies van der Rohe's 1958 Seagram Building in New York (fig. 7). In the "602 A" radio, the grid also symbolises modernism while its

Die horizontalen Linien und die übereinander angebrachten, herausstehenden Scheibenstreifen der Radios erinnern einerseits an die Decks eines Dampfers. Andererseits könnten sie auf das ebenfalls an Dampfergestaltung angelehnte Architekturmotiv der ‚Augenbrauen' über Fenstern verweisen, wie es sich etwa bei Bauten von Frank Lloyd Wright, Art Déco-Bauten oder der Streamline-Moderne findet.[42] In der Architektur findet sich ebenfalls das Motiv der Bullaugen-Reihe, das hier von den Reglerknöpfen unter den ‚Decks' repräsentiert wird.[43]

Im Gegensatz zu den vorhergehenden Modellen zeigt das Radiomodell „6 D 030 Z" der Evans Plywood Company für die Firma Zenith (Kat.Nr. 42) den pointierten Einsatz von traditionellen Architekturelementen. Designt haben es 1946 der Architekt Charles Eames und seine Frau, die Bildhauerin Ray Eames. Das klassische Bauelement des Rundbogens ist auf den Kopf gestellt und sein keilförmiger ‚Schlussstein' unterteilt hier die Bogenfläche. Er wird zum Pfeil, der auf den Reglerknopf zeigt. Ein vergleichbarer Umgang mit traditionellen Bauelementen ist in der Architektur des 20. Jahrhunderts immer wieder zu beobachten. Die von ihrer ursprünglichen Funktion gelösten Bauelemente erhalten einen eher zeichenhaften Charakter, der zitathaft auf Geschichtlichkeit verweist, mit ihr aber spielerisch umgeht.[44]

Eine typische Fensterlösung der Moderne nutzt dagegen das Modell „602 A" der US-Firma Emerson Radio and Phonograph Cooperation, das 1949 Raymond Loewy entwarf (Kat.Nr. 60). Mit seiner kräftigen Form und dem mit tiefen Öffnungen versehenen Rastergitter vor dem Lautsprecher erscheint es wie die Modellversion eines Baus mit Rasterfassade. Diese Art Bauten beherrschen die Stadtbilder moderner Großstädte wie beispielsweise Ludwig Mies van der Rohes Seagram Building von 1958 in New York (Abb. 7). Bei diesem Modell steht das Raster zwar auch für Modernität, aber die Schwere seiner Ausführung, die Symmetrie des Gerätes sowie die Goldfarbe und das Weinrot des Materials wirken traditionsverbunden. Ähnliche Tendenzen – wie etwa goldeloxiertes Aluminium für Fensterrahmen – lassen sich an einer Vielzahl von Bauten der 1950er Jahre wiederfinden, die zwischen Modernität und Tradition changieren.

Rang, Wert, Magie

Eine Verbindung von Architektur und Radio stellt auch die deutsche Objektkünstlerin Isa Genzken her. In ihrer Arbeit „Deutsche Bank Proposal" aus dem Jahr 2000 setzt sie einem

bulky form, its symmetrical design and the golden and claret colours of the material have a traditional feel. Similar features, for example window frames made of gold-anodised aluminium, can be found in many buildings from the 1950s, which oscillate between modernism and tradition.

Status, Value, Magic

German artist Isa Genzken also creates a connection between architecture and radio design: in her piece 'Deutsche Bank Proposal' from the year 2000 she attaches two telescopic aerials to a model of the Sony Building in New York (formerly the AT & T Building by Philip Johnson, 1979-84), which seems only logical for the building of a radio manufacturer. In contrast to the radios presented here, in this predecessor of the postmodern high-rise with its pediment reminiscent of a cabinet, the architecture virtually becomes the radio.

The radios introduced here are not architectural models of existing or planned buildings. Only the "66" radio comes close to a design that could actually be built. The other radios only feature similarities to façade design or to flat roofs or to specific architectural elements. In particular, the central element of the loudspeaker has inspired designers to come up with some very creative solutions.[45] Due to their function of connecting the interior with the exterior, windows or other architectural openings offer themselves as rather obvious references to highlight the radio's main function of distributing sound: just like architecture, the radio is a sound space or sound volume.

Through the use of architectural references, radio design exploits the high status of architecture. With their traditional architectural forms, the

Modell des Sony Buildings in New York (ehem. AT & T Building, Philip Johnson, 1979-84) zwei Teleskopantennen auf. Für das Hochhaus eines Herstellers von Unterhaltungselektronik erscheint das nur sinnvoll. Im Gegensatz zu den hier vorgestellten Radios wird damit bei der Inkunabel des postmodernen Hochhausbaus, mit ihrem an ein Schrankmöbel erinnernden Giebelaufsatz, quasi die Architektur selbst zum Radio.

Die vorgestellten Radios sind dagegen keine Architekturmodelle von existierenden oder geplanten Bauten. Einer ‚Baubarkeit' kommt unter ihnen nur Modell „66" nahe. Bei allen anderen sind entweder nur Ähnlichkeiten zum Fassadenaufbau, gegebenenfalls dem Flachdach oder mit Baudetails erkennbar. Dabei hat das zentrale Element des Lautsprechers die Gestalter zu besonders kreativen Lösungen herausgefordert.[45] Wegen ihrer vergleichbaren Funktion Innen und Außen zu verbinden, bieten sich Verweise auf Fenster oder andere architektonische Öffnungen an. Damit wird die zentrale Funktion des Radios, Klang zu verbreiten, unterstrichen. Wie auch die Architektur ist das Radio ein Klangraum beziehungsweise Klangkörper.

Die Gestaltung der Radiogeräte nutzt durch die Architekturbezüge den hohen Rang der Baukunst. Durch traditionelle Architekturformen passen sich die Radios herkömmlichen Wohnansprüchen an und durch zeitgenössische Aspekte der Architektur wirken sie modern. Wichtig ist dabei, dass die Geräte zumindest als repräsentative oder luxuriöse, wenn nicht sogar pseudoreligiöse Objekte erscheinen. So werden Wert und Magie des Radios ausgespielt – eben Luxusleben und Gottesdienst!

1   Vgl. Friemert, 1996, S. 14.
2   Vgl. Kölsch, 1990, S. 5.
3   Vgl. Ebd., S. 8.
4   Vgl. Ebd., S. 4-6 und Wichmann, 1993, S. 48.
5   Vgl. Ebd., S. 6.
6   Vgl. Kölsch, 1990, S. 3.
7   Vgl. Schmidt, 1991, S. 76.
8   Vgl. Wichmann, 1993, S. 48-49.
9   Z.B. St. Engelbert, Köln-Riehl, 1930, D. Böhm.
10   Vgl. Baumerich, 2005, S. 153.
11   Vgl. Kölsch, 1990, S. 5.
12   Vgl. Christie, 2006, S.382.

radios shown here conform to traditional ideas of living while using contemporary architectural aspects to project a sense of modernity. Another important aspect is that the devices look at least representative or luxurious, if not pseudo-religious. The value and magic of the radio are being fully exploited: luxury and worship!

1   See Friemert, 1996: 14.
2   See Kölsch, 1990: 5.
3   Ibid.: 8.
4   Ibid.: 4-6 and Wichmann, 1993: 48.
5   See Wichmann, 1993: 6.
6   See Kölsch, 1990: S. 3.
7   See Schmidt, 1991: 76.
8   See Wichmann, 1993: 48-49.
9   For example: St. Engelbert, Köln-Riehl, 1930, D. Böhm.
10   See Baumerich, 2005: 153.
11   See Kölsch, 1990: 5.
12   See Christie, 2006: 382.
13   See Kölsch, 1990: 5.
14   For example: Thomas Jefferson's Monticello country estate, 1768-1809.
15   See Kölsch, 1990: 9.
16   The height of the Empire State Building in New York (1930/31), designed by William F. Lamb, was only surpassed in 1972 by the World Trade Center (1970/71), also located in New York and designed by Minoru Yamasaki.
17   See Schmidt, 1991: 66-67.
18   Attwood, 1997: 8.
19   See Johnson, 2000: 46.
20   See Sioux City Municipal Auditorium, Sioux City, Knute E. Westerlind, 1938-1950.
21   See Bresford Hotel, Glasgow, James W. Weddel, 1938, or Odeon Cinema, Brighton, Andrew Mather, 1937.
22   See Eastern Columbia Building, Los Angeles, Claude Beelman, 1930. On art déco design in radios see Johnson/Johnson, 1989: 87.
23   See Attwood, 1997: 12.
24   See Eastern Columbia Building, Los Angeles, Claude Beelman, 1930.

13 Vgl. Kölsch, 1990, S. 5.

14 Z.B. Thomas Jeffersons Landgut Monticello, 1768-1809.

15 Vgl. Kölsch, 1990, S. 9.

16 So wurde das Empire State Building in New York von 1930/31, William F. Lamb an Höhe erst 1970/71 durch das Minoru Yamasaki und 1972 durch das World Trade Center, beide New York, übertroffen.

17 Vgl. Schmidt, 1991, S. 66-67.

18 Attwood, 1997, S. 8.

19 Vgl. Johnson, 2000, S. 46.

20 Vgl. Sioux City Municipal Auditorium, Sioux City, Knute E. Westerlind, 1938-1950.

21 Vgl. Bresford Hotel, Glasgow, James W. Weddel, 1938, oder Kino Odeon, Brighton, Andrew Mather, 1937.

22 Vgl. Eastern Columbia Building, Los Angeles, Claude Beelman, 1930. Zu Art Déco bei Radios vgl. Johnson/Johnson, 1989, S. 87.

23 Vgl. Attwood, 1997, S. 12.

24 Vgl. Eastern Columbia Building, Los Angeles, Claude Beelman, 1930.

25 Z.B. Pavillon L'Hôtel du Collectionneur, Pierre Patout, 1925; Empire State Building, New York, William F. Lamb, 1930/31.

26 Vgl. Schmidt, 1991, S. 64.

27 Z.B. Rathaus, Los Angeles, John C. Austin, John Parkinson, Albert C. Martin, 1926-28.

28 Schmidt, 1991, S. 65.

29 Z.B. Maybury Roadhouse, Edinburgh, Patterson & Broom, 1936.

30 Z.B. Chrysler Building, New York, William Van Alen, 1928-1931.

31 Z.B. Rathaus. Los Angeles, John C. Austin, John Parkinson, Albert C. Martin, 1926-28.

32 Vgl. Attwood, 1997, S. 18.

33 Vgl. Schmidt, 1991, S. 19.

34 Z.B. Art Deco: 1200 Collins Avenue, Miami Beach, L. Murray Dixon, 1939; ‚Bauhaus-Moderne': Haus Tugendhat, Brünn, Ludwig Mies van der Rohe, 1928.

35 Z.B. ‚Bauhaus-Moderne': Wohnheim auf der Werkbund-Ausstellung, Breslau, 1928/29, Hans Scharoun; Konstruktivismus: Gebäude der Iswestija, Moskau, 1925/26, Grigori Borissowitsch Barchin.

36 Vgl. Kölsch, 1990, S. 3.

37 Art Deco: 1200 Collins Avenue, Miami Beach, L. Murray Dixon, 1939; ‚Bauhaus-Moderne': Haus Tugendhat, Brünn, Ludwig Mies van der Rohe, 1928.

38 Vgl. Johnson/Johnson, 1989, S. 88.

39 Vgl. Battersby, 1976, S.111-119 und Colloway, 1991, S. 276-277.

40 Z.B. Bauhaus-Gebäude, Dessau, Walter Gropius, 1925-1926. Eastern Columbia Building, Los Angeles, Claude Beelman, 1930.

41 Vgl. Attwood, 1997, S. 17.

42 Z.B. Fallingwater-Haus, Bear Run, Frank Lloyd Wright, 1935-39, und 1200 Collins Avenue, Miami Beach, L. Murray Dixon, 1939.

43 Z. B. Hotel Park Central, Miami Beach, Henry Hohauser, 1937.

44 Z. B. Kaufhaus Tietz, Bamberg, Johannes Kronfuß, 1909-1910.

45 Vgl. Kölsch, 1990, S. 9.

25 For example: Pavillon L'Hôtel du Collectionneur, Pierre Patout, 1925; Empire State Building, New York, William F. Lamb, 1930/31.

26 See Schmidt, 1991: 64.

27 For example: City Hall, Los Angeles, John C. Austin, John Parkinson, Albert C. Martin, 1926-28.

28 Schmidt, 1991: 65.

29 For example: Maybury Roadhouse, Edinburgh, Patterson & Broom, 1936.

30 For example: Chrysler Building, New York, William Van Alen, 1928-1931.

31 For example: City Hall. Los Angeles, John C. Austin, John Parkinson, Albert C. Martin, 1926-28.

32 See Attwood, 1997: 18.

33 See Schmidt, 1991: 19.

34 For example: Art Deco: 1200 Collins Avenue, Miami Beach, L. Murray Dixon, 1939; 'Bauhaus-Modernism': Haus Tugendhat, Brünn, Ludwig Mies van der Rohe, 1928.

35 For example: 'Bauhaus-Modernism': Residence for Singles at Werkbund Exhibition, Breslau, 1928/29, Hans Scharoun; Constructivism: Iswestia Building, Moscow, 1925/26, Grigori Borissowitsch Barchin.

36 See Kölsch, 1990: 3.

37 Art Deco: 1200 Collins Avenue, Miami Beach, L. Murray Dixon, 1939; 'Bauhaus-Modernism': Haus Tugendhat, Brünn, Ludwig Mies van der Rohe, 1928.

38 See Johnson/Johnson, 1989: 88.

39 See Battersby, 1976: 111-119 and Colloway, 1991: 276-277.

40 For example: Bauhaus-Building, Dessau, Walter Gropius, 1925-1926. Eastern Columbia Building, Los Angeles, Claude Beelman, 1930.

41 See Attwood, 1997: 17.

42 For example: Fallingwater-House, Bear Run, Frank Lloyd Wright, 1935-39, und 1200 Collins Aenue, Miami Beach, L. Murray Dixon, 1937.

43 For example: Hotel Park Central, Miami Beach, Henry Hohauser, 1937.

44 For example: Kaufhaus Tietz, Bamberg, Johannes Kronfuß, 1909-1910.

45 See Kölsch, 1990: 9.

Walter Maria Kersting
Pionier des Kunststoffdesigns

Walter Maria Kersting
Pioneer of Plastic Design

Günter Lattermann

Kersting? Wer war das? Was hat er gemacht? Mit diesen Fragen könnte man sich beschäftigen, wenn man etwas über frühe Designgeschichte wissen möchte. Walter Maria Kersting war schon vor dem Zweiten Weltkrieg auf vielen Gebieten tätig, wie Architektur, Innenarchitektur, Grafik-, Industrie- und Kunststoffdesign.[1]

Meist taucht aber der Name Walter Maria Kersting nur im Zusammenhang mit dem berühmt-berüchtigten „Volksempfänger" auf. Um diese eindimensionale Betrachtung etwas aufzubrechen, soll an dieser Stelle auf seine Aktivitäten als einer der Pioniere des Kunststoffdesigns in Deutschland eingegangen werden.

Zunächst zur Biografie Walter Maria Kerstings (Abb. 1):[2] Er wurde am 8. Juli 1892 in Münster in Westfalen als Sohn eines Architekten geboren und verstarb am 5. Mai 1970 in Waging am See, Oberbayern. Er studierte von 1912 bis 1914 Maschinenbau und Architektur an der TH Hannover und war ab 1914 als grafischer Gestalter in Hannover unter anderem für die Firma Bahlsen tätig. Nach Kriegsende, ab 1918, vergrößerte er sein grafisches Atelier in Hannover und gründete den Bund Hannoverscher Druckgraphiker, aus dem 1920 der Bund Deutscher Gebrauchsgraphiker hervorging. Als künstlerischer Leiter einer großen Druckerei verbrachte er die Jahre 1921 bis 1922 in São Paulo, Brasilien. Nach seiner Rückkehr aus gesundheitlichen Gründen arbeitete er von 1922 bis 1927 als Architekt und Grafiker in Weimar, anschließend unter anderem für die DKW Autowerke in Zschopau, aber auch als künstlerischer Leiter und Architekt bei den Krauss-Werken (Waschmaschinen) im sächsischen Schwarzenberg. Von 1927 bis 1932 hatte er auf Vorschlag Richard Riemerschmids[3] eine Professur für Kunsterziehung, Grafik und Industrie-Modelle an den Kölner Werkschulen inne. Dies war der früheste Versuch in Deutschland überhaupt, eine Abteilung für industrielle Formgestaltung einzurichten. Laut Aussage seines Sohnes Arno bot ihm Walter Gropius 1928 eine Stellung am Bauhaus an. Allerdings habe er dieses Angebot, nach Dessau zu wechseln, nicht wahrgenommen, da er von den am Bauhaus kursierenden marxistischen Ideen, besonders kurz vor der Leitungsübernahme durch Hannes Meyer, der diese Tendenzen stark förderte, nicht überzeugt gewesen sei.[4] Nachdem er dann 1932, zusammen mit anderen Lehrkräften, die Kölner Werkschule aus Protest gegen eine drastische Gehälter-Kürzung um die Hälf-

Kersting? Who was that? What did he do? If you wanted to learn more about early design history, these would be some of the questions to ask. As early as before the Second World War, Walter Maria Kersting had already worked in many different fields such as architecture, interior design, graphic design, industrial design and plastic design.[1]

However, his name appears mainly in connection with the notorious 'Volksempfänger' radio. In order to rectify this rather one-dimensional view of his work, this essay will look at Kersting's activities as one of the pioneers of plastic design in Germany.

1 / Porträt / Portrait, Walter Maria Kersting, ca. 1930, Nachlass / Estate Walter Maria Kersting.

te (durch den damaligen Oberbürgermeister Konrad Adenauer) verlassen hatte, war er von 1932 bis 1933 als selbstständiger Formgeber und Grafiker in Berlin tätig. Dort arbeitete er unter anderem an der Gestaltung von Elektroartikeln für die AEG.[5] Vielleicht begünstigt durch seine kritische Haltung gegenüber dem Bauhaus, trat er ab 1933 eine Professur für Angewandte Grafik an der Kunstakademie Düsseldorf an, die er, obwohl er persönlich „ein unversöhnlicher Gegner der Nazis war"[6], bis 1943 innehatte. Nachdem 1942 sein Wohnhaus teilweise ausgebombt und die Kunstakademie zerstört worden war, wurde er wegen Spionageverdacht und politischer Unzuverlässigkeit 1943 entlassen.[7] 1945 baute er mit seinen Söhnen eigene Werkstätten für technische Gestaltung in Waging am See auf und richtete nach dem Krieg zusätzlich noch ein Büro für Formgebung in Düsseldorf ein. Nachdem er vorschlug den „Rat für Formgebung" zu gründen, gehörte er diesem seit 1951 als Gründungsmitglied an. Er empfahl bereits 1950/51 in einem detaillierten Programm den Aufbau einer „Hochschule für Industriegestalter/Hochschule für Industrielle Formgebung" unter anderem in Hannover. Weiterhin regte er auf der Hannover-Messe eine Sonderschau „Industrielle Formgebung" an.

Wie aus Originalentwürfen in seinem Nachlass hervorgeht, hatte Kersting 1932 verschiedene Radiomodelle aus Phenolharz für die AEG entworfen. Als 1933 die deutsche Elektroindustrie für den Bau des Volksempfängers „VE 301" zusammengetrommelt wurde,[8] um verschiedene Planungen für die preiswerteste Technik und das ansprechendste Gehäuse-Design einzuholen, reichte neben anderen auch Walter Maria Kersting Pläne für die Gestaltung des Pressstoff-Gehäuses ein (Abb. 2). Laut Aussagen seiner Söhne Arno und Gerwald,[9] die ihrerseits Designer waren, teilweise von ihm ausgebildet wurden und nach dem Zweiten Weltkrieg eng mit ihrem Vater in den Waginger Werkstätten zusammenarbeiteten, stammte der Kersting-Entwurf für den ersten Volksempfänger also nicht von 1928, wie in der Biografiebroschüre angegeben.[10] Kerstings Entwurf stimmte in den Hauptmerkmalen mit dem bekannten Gehäuse des ersten Volksempfängers „VE 301 W" überein, unterschied sich jedoch in bestimmten Details. So fehlte zum Beispiel der Rundwulst an den Gehäuserändern. Weiterhin war der Lautsprecherrand nicht glatt, sondern nach innen konisch geriffelt. Die Knöpfe im Entwurf waren weiß und das Emblem des „Rufenden Adlers" im Rundpor-

Let's start with Walter Maria Kersting's biography (fig. 1):[2] The son of an architect, he was born on July 8th, 1892 in the Westphalian town of Münster and he died on May 5th, 1970 in Waging am See in Upper Bavaria. From 1912 to 1914, he studied mechanical engineering and architecture at the Technical University of Hanover and, from 1914 onwards, he worked as a graphic designer in Hanover, for example for the Bahlsen company. After the war, in 1918, he extended his graphic design studio in Hanover and founded the Bund Hannoverscher Druckgraphiker (Association of Hanover Print Designers), which, in 1920, became the Bund Deutscher Gebrauchsgraphiker (German Association of Applied Graphics Designers). From 1921 to 1922, he lived in São Paulo, Brazil, where he was the artistic director of a large printing house. For health reasons, he returned to Germany, working as an architect in Weimar from 1922 to 1927, followed, among other positions, by a job at car manufacturer DKW in Zschopau and employment as architect and artistic director at Krauss-Werke (washing machines) in the Saxon town of Schwarzenberg. Following a recommendation by Richard Riemerschmid[3], from 1927 to 1932, he was professor of art education, graphic design and industrial design at the Kölner Werkschulen. This represented the first attempt in Germany to establish a dedicated department for industrial design. According to Kersting's son Arno, in 1928, Walter Gropius offered Kersting a position at the Bauhaus. However, Kersting declined the opportunity to move to Dessau because he was not convinced by the Marxist ideas which were circulating at the Bauhaus at the time, especially shortly before Hannes Meyer, who strongly supported these ideas, took over as director.[4] In protest against a drastic 50% cut in wages (enforced by the then-mayor Konrad Adenauer), he left the

tal an der Vorderfront schräg gestellt.[11] Dieses Emblem stammt übrigens nicht von Kersting selbst. Auf Plakaten mit dem Adlerkopf und den kreisförmigen Radiowellen findet sich gelegentlich der Namenszug „Riemer". Der Gehäuseentwurf Walter Maria Kerstings wurde in bestimmten Details eigenmächtig abgeändert und als Gemeinschaftserzeugnis von vielen Presswerken hergestellt (Kat.Nr. 5 und 6) und mit dem Chassis aus den insgesamt 28 in Deutschland bestehenden Apparatefabriken versehen.[12]

2 / Entwurf / Draft, „Volksempfänger", Walter Maria Kersting (DE), 1933, Nachlass / Estate Waler Maria Kersting.

Arno Kersting berichtet, der Vater sei mit den seiner Meinung nach schlechten Änderungen nicht einverstanden gewesen und hätte sich so sehr darüber geärgert, dass er keine Verantwortung dafür übernehmen wollte. Offenbar erhielt er auch

Kölner Werkschulen in 1932 and went to Berlin, where, from 1932 to 1933, he was an independent industrial and graphic designer. In Berlin, he designed, among other things, electrical appliances for AEG.[5] His critical attitude towards the Bauhaus may have played a role in his becoming professor of applied graphic design at the Düsseldorf Academy of the Arts in 1933. Although he was "irreconcilably opposed to the Nazis"[6], he held this position until 1943. After both his home and the academy had been (partly) destroyed in the 1942 bombings, in 1943, he was dismissed on charges of suspected espionage and political unreliability.[7] Together with his sons, in 1945, he established his own studios for technical design in Waging am See and he also set up a studio for industrial design in Düsseldorf. The Rat für Formgebung (German Design Council) was founded on his initiative, and he was a member of the council until 1951. As early as 1950/51, he developed a detailed programme for establishing a university of industrial design, in, among other places, Hanover. He also initiated a special exhibition of industrial design at the Hanover trade fair.

Original designs included in his estate show that, in 1932, Kersting designed different radio devices for AEG with casings made from phenolic resin. When, in 1933, the German electrical appliances industry was called upon to submit designs for the "VE 301 Volksempfänger"[8] in order to find the most cost-efficient technology and the most appealing design, Walter Maria Kersting was among those who submitted plans for the design of the moulded casing (fig. 2). According to his sons, Arno and Gerwald,[9] – who were designers themselves, had partly been trained by their father and had closely collaborated with him at the Waging studios – the Kersting design for the first Volksempfänger

keinen weiteren Auftrag für die endgültige Gestaltung. Dies und die Abneigung Kerstings gegenüber den Nazis schlossen ihn für den gewünschten Propagandarummel aus. Kersting wurde nie als Gestalter des für die Verbraucherakzeptanz so wichtigen Gehäuses genannt. Lediglich der stramme Parteigenosse Otto Grießing (Konstruktionsaufsicht der gesamten Technik) stand immer als alleiniger ‚Vater des Volkempfängers' im Vordergrund.[13]

Insgesamt scheint Kersting mit dieser „Ikone des Designs"[14] „eine ästhetisch befriedigende und funktional übersichtliche Produktform mit einer Spur Art Déco [...] [und einer] Prise Bauhaus-Sachlichkeit"[15] aber im Rückblick eher negativ belastet worden zu sein: „Die klare einfache Form und das Material Bakelit ermöglichten seine industrielle Massenfertigung und einen niedrigen Preis [...]. Walter Maria Kerstings Design für den in Massen produzierten Volksempfänger ist ein bizarres Paradox [...]. Dieses Radio [...] manipulierte als Sprachrohr der Machthaber das kollektive Bewusstsein und trug zur Unterminierung der kritischen und intellektuellen deutschen Tradition bei"[16].

Aus Walter Maria Kerstings Nachlass stammen weitere Gestaltungen von Radiogehäusen aus Press-Phenolharz für die AEG, von denen einer 1932 entworfen und als „Geadux 112" produziert wurde (Abb. 3 und 4).[17] Kersting arbeite noch mindestens an zwei weiteren Elektrogeräten der AEG. So entwarf er um 1928

4 / Geadux 112, Walter Maria Kersting, AEG, Berlin (DE) 1933-34. Sammlung / Collection Lattermann © Foto / Photo: Rheinisches Bildarchiv Köln / Cologne, Marion Mennicken.

did not date from 1928 as stated in the biography brochure.[10] In its main characteristics, Kersting's design was consistent with the well-known casing of the first Volksempfänger, the "VE 301 W", but it differed in its detailing. For example, the casing did not have the inset bevelling around the front edges. Furthermore, the bezel around the speaker was not smooth; instead its inner surface featured conical grooves. Also, the controls were white and the emblem of the 'calling eagle' in the rounded arch at the front was set at an angle.[11] The emblem, however, was not designed by Kersting. Some of the posters featuring the eagle's head and circular radio waves

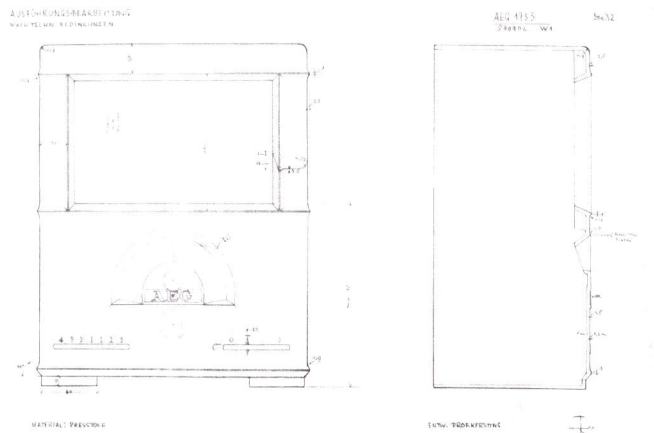

3 / Entwurf / Draft, Walter Maria Kersting, (DE) 1932. Nachlass / Estate Walter Maria Kersting, © Foto / Photo: Rheinisches Bildarchiv Köln / Cologne, Marion Mennicken.

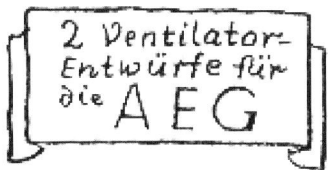

2 Ventilator-
Entwürfe für
die AEG

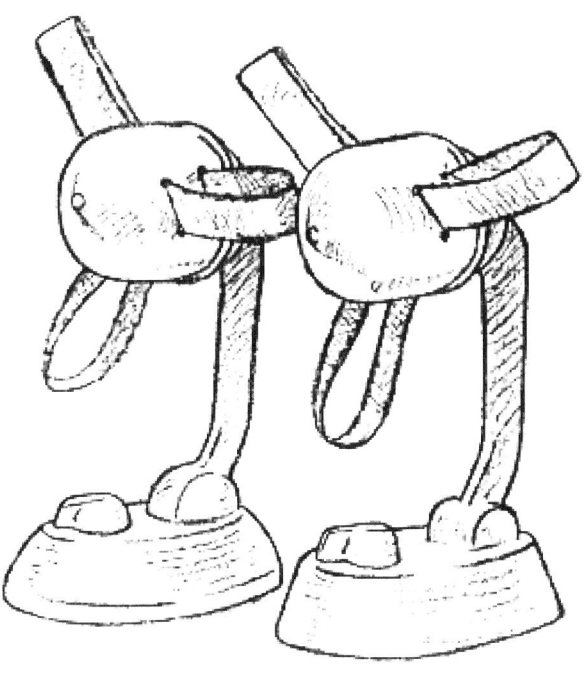

5 / Entwurf / Draft, Walter Maria Kersting (DE), 1933,
Nachlass / Estate Walter Maria Kersting.

laut Aussage seines Sohnes Arno das Gehäuse aus Press-Phenolharz des Staubsaugers „Vampyr 100".[18] Weiterhin gestaltete er 1933
die alte Version des Bandventilators der AEG neu (Abb. 5 und 6).[19]
Sowohl die alte (englisch: Bandolero), als auch die neue Version
(englisch: Ribbonaire) wurden in Großbritannien (Firma Ekco)
und den USA (Firmen Diehl und Singer) offensichtlich in Lizenz
produziert.

are signed with the name 'Riemer'. Details of Walter
Maria Kersting's casing design were changed without his permission and the new design was manufactured as a co-designed product by many different moulding companies (cat.no. 5 and 6) and fitted
with the chassis made in the 28 German appliances factories.[12] According to Arno Kersting, his father
did not approve of the changes. He regarded them
as retrograde and was so angry about these changes
that he refused to take responsibility for them. Apparently, he was not commissioned with the final
design. Because of this rift and due to Kersting's dis-
like of the Nazis, he was no longer useful to the de-
sired propaganda campaign. Kersting was never
mentioned as the designer of the casing, the part of
a device that plays such an important role in consumer acceptance. Only staunch party member Otto
Grießing (head of technological construction) was
named as the sole 'father of the Volksempfänger'.[13]

In retrospect however, the Volksempfänger
'design icon'[14] with its "aesthetically satisfying and
functionally clear product form, featuring a touch of
art déco [...] [and a] pinch of Bauhaus objectivity",[15]
seems to have cast a shadow on Kersting's reputa-
tion: "Its clear, simple form together with the use of
Bakelite allow industrial production and a low price
[...]. Walter Maria Kersting's design for the mass-produced Volksempfänger is a bizarre paradox [...]. As
the ruling powers' mouthpiece, this radio [...] ma-
nipulated the collective consciousness and contributed to undermining the critical and intellectual
German tradition."[16]

In Walter Maria Kersting's estate there are also
other designs of radio casings made from pressed
phenolic resin, which he developed for AEG. One
of these designs was created in 1932 and produced

Walter Maria Kersting, der schon 1932 den ‚Ingenieur-Künstler' gefordert hatte,[20] meinte: „Die wirkliche industrielle Formgebung setzt außer künstlerischen Fähigkeiten sehr weitgehendes technisches Ingenieurkönnen voraus"[21]. Er war nicht nur selbst der Altmeister der Industrie-Designer, sondern zudem auch einer der (vergessenen) Pioniere des Deutschen Kunststoffdesigns seit Anfang der 1930er Jahre.[22] Er hat die Designgeschichte ganz wesentlich geprägt. Es lohnt sich, Walter Maria Kersting wieder zu entdecken.

1   Vgl. Krauss, 1933; Kersting, 1928; Kersting 1932; Kersting, 1933; Kersting: Kersting Marken, 1955; Kersting: Technische Gestaltung, 1955; Pfaender/Baum/Schäfer, 1974; Altmayer 2013.

2   Informationen aus dem Nachlass Walter Maria Kersting und aus persönlichen Mitteilungen des Sohnes Arno Kersting an den Autor im Jahr 2002.

3   1926-1931: Direktor der Kölner Werkschulen

4   Persönliche Mitteilungen des Sohnes Arno Kersting an den Autor im Jahr 2002.

5   Aus dem Nachlass Walter Maria Kerstings.

6   Handschriftliche Erklärung des zu den entarteten Künstlern zählenden Otto Pankok vom 28. Februar 1947, Personalakte Walter Maria Kersting BR-PE Nr. 7604, Nordrhein-Westfälisches Hauptstaatsarchiv Düsseldorf.

7   Aus dem Brief des Abschnittleiters der NSDAP Brandner vom 30.08.1944: „Kersting war schon einmal in Haft. In seinem gesamten Verhalten ist er als nicht zuverlässig zu bezeichnen", in: Personalakte Walter Maria Kersting BR-PE Nr. 7604, Nordrhein-Westfälisches Hauptstaatsarchiv Düsseldorf.

8   Vgl. König, 2003.

9   Persönliche Mitteilungen der Söhne Arno und Gerwald Kersting an den Autor im Jahr 2002.

10   Vgl. Pfaender/Baum/Schäfer, 1974.

11   Vgl. Ebenda.

12   Vgl. König, 2003.

13   Lattermann, 2004, S. 940.

14   Ebenda.

15   Selle, 2007, S. 200-202.

16   Kras, 2000, S. 56-57.

17   Aus dem Nachlass Walter Maria Kerstings.

18   Persönliche Mitteilung des Sohnes Arno Kersting an den Autor im Jahr 2006.

19   Vgl. Sonderdruck aus den AEG-Mitteilungen, Heft 2, 1933, Archiv Deutsches Technikmuseum Berlin und vgl. [o.A.]: Der Bandventilator, 1933/1934, S. 118.

20   Kersting, 1932.

21   Kersting: Über den Formgeber, 1955.

22   Lattermann, 2006.

as "Geadux 112" (fig. 3 and 4).[17] Kersting worked on at least two more electrical appliances for AEG. According to his son Arno, he designed the pressed phenolic resin casing for the "Vampyr 100" vacuum cleaner.[18] In 1933, he also redesigned the old version of the "Bandolero" from AEG (fig. 5 and 6).[19] Apparently, both the old and the new "Ribbonaire" version were produced under licence in Great Britain (by Ekco) and in the USA (by Diehl and Singer).

Walter Maria Kersting, who advocated the 'engineer-artist' as early as 1932,[20] stated: "Besides artistic skills, true industrial design requires very broad skills in technical engineering."[21] Kersting was not only an early master of industrial design, but also one of the (forgotten) pioneers of German plastic design since the 1930s.[22] He has significantly shaped design history. It is therefore worthwhile to rediscover the work of Walter Maria Kersting.

1   See Krauss, 1933; Kersting, 1928; Kersting 1932; Kersting, 1933; Kersting: Kersting Marken, 1955; Kersting: Technische Gestaltung, 1955; Pfaender/Baum/Schäfer, 1974 and Altmayer 2013.

2   Information from Walter Maria Kersting's estate and from personal correspondence by his son Arno Kersting to the author in 2002.

3   1926-1931: director of Kölner Werkschulen.

4   Personal correspondence from his son Arno Kersting to the author in 2002.

5   From Walter Maria Kersting's estate.

6   Handwritten declaration by Otto Pankok, who was regarded as one of the degenerate artists, dating from 28 February 1947; Walter Maria Kersting personal file BR-PE Nr. 7604, Nordrhein-Westfälisches Hauptstaatsarchiv Düsseldorf.

7   From a letter by NSDAP section leader Brandner dated 30 August 1944: "Kersting was arrested once before. His overall attitude must be regarded as unreliable." in: Walter Maria Kersting personal file BR-PE Nr. 7604, Nordrhein-Westfälisches Hauptstaatsarchiv Düsseldorf.

8   See König, 2003.

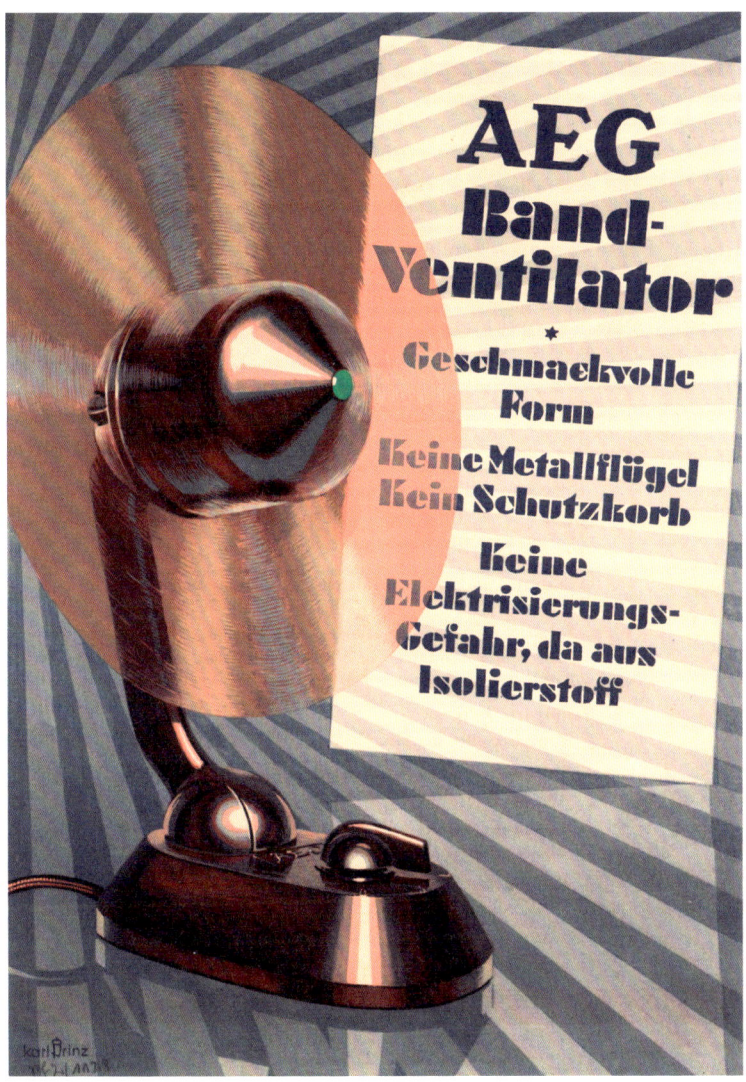

6 / Reklame / Advertisement, „Bandventilator",
Walter Maria Kersting, AEG, Berlin (DE) 1928,
Stiftung Deutsches Technikmuseum Berlin
© Foto / Photo: Historisches Archiv.

9   Personal correspondence from Kersting's sons, Arno and
    Gerwald Kersting, to the author in 2002.
10 See Pfaender/Baum/Schäfer, 1974.
11 Ibid.
12 See König, 2003.
13 Lattermann, 2004: 940.
14 Ibid.
15 Selle, 2007: 200-202.
16 Kras, 2000: 56-57.
17 From Walter Maria Kersting's estate.
18 Personal correspondence from Kersting's son Arno Kersting to
    the author in 2006.
19 See special reprint from AEG-Mittelungen, issue 2, 1933,
    Archiv Deutsches Technikmuseum Berlin and [no author]: Der
    Bandventilator, 1933/1934: 118.
20 Kersting, 1932.
21 Kersting: Über den Formgeber, 1955.
22 Lattermann, 2006.

Der
Matrizen-Stil

The Matrix Style

Günter Lattermann

Der Ausdruck „Matrizen-Stil" wurde 1993 von dem Architekten und Kunststoffsammler Hans Ulrich Kölsch für eine spezielle Gestaltungsrichtung geprägt, die sich vom späten Jugendstil bis ins Art Déco, das heißt vom ersten bis zum vierten Jahrzehnt des 20. Jahrhunderts erstreckt.[1] Die Besonderheit dieses Stils lag in seiner Bezugnahme auf eine materielle Revolution, nämlich der Herstellung des ersten synthetischen Kunststoffs Phenolharz (Handelsname Bakelit) ab 1910 und seiner Formgebung zu Massenobjekten der Industrie und des täglichen Bedarfs. Bis zum Aufkommen der späteren, thermoplastischen Kunststoffe wurden diese ersten Materialien (Duromere, Duroplaste) durch ein Pressverfahren erstmals in großer Serie hergestellt. Phenolharze (‚Das Material der tausend Möglichkeiten') waren somit für ihre vielfältigen Einsatzgebiete in großer Zahl zugänglich – ein Phänomen, dass zwar Vorgänger hatte (zum Beispiel Bois Durci und Papiermaché-Artikel[2]), aber seine umfassende Bedeutung erst mit den neuen synthetischen Materialien gewann. Dies erstreckte sich auch auf die neu aufkommenden Fragen der Gestaltung solcher industriellen Massenware. Hans Ulrich Kölsch schreibt dazu, dass es wahrscheinlich ein Zufall sei, dass das entscheidende Phenolharz-Patent Leo Hendrik Baekelands von 1907 und das Gründungsjahr des Deutschen Werkbundes zusammenfielen. Es scheine aber symptomatisch für eine Zeit ständiger Konfrontation mit Neuerungen. Die ehrenwerten und berühmten Herren dieser Vereinigung von Architekten, Künstlern und Industriellen hätten es sich zur Aufgabe gemacht, reformerisch und aufklärend auf alle Gebiete von Wissenschaft und Kunst einzuwirken. Der Jugendstil als Protest gegen historisierenden Schwulst schien überwunden – oder sei zumindest auf dem Rückzug gewesen. Er hätte das längst installierte Maschinenzeitalter nur kurz mit seinen zarten Blüten umrankt. Das neue Jahrhundert hätte neue Aufgaben gestellt: Der enorme Bevölkerungszuwachs machte einfachere, radikalere Lösungen für soziale Probleme notwendig. Verkehr und Produktion von Gebrauchsgütern mussten ebenso radikal folgen.[3]

Die Nähe zur industriellen Produktion erzwang eine Form, die mit den technischen Notwendigkeiten und der Funktion eines Gegenstandes in Einklang stand.[4] Dies geschah nunmehr in großem Ausmaß mit den neuen synthetischen Kunststoffen (Phenolharze und Harnstoffharze) mit ihren charakteristischen Verarbeitungsverfahren, dem Pressen unter Hitze (Phenolharze bei 140-180° C, Harnstoffharze bei 140-145° C) und bei hohem Druck.[5]

The term 'matrix style' was coined in 1993 by architect and plastic design collector Hans Ulrich Kölsch. It designates a particular design style, spanning from late art nouveau to art déco, that is, from the first to the fourth decade of the 20th century.[1] What was special about this style was that it was linked to a revolutionary material – phenolic resin (trade name Bakelite), the first synthetic plastic produced in 1910 – used in the design of industrial or consumer products. Using compression moulding, these early synthetic materials (duromers, thermosets) were, for the first time, produced in large quantities until they were later replaced by thermoplastics. A large number of phenolic resins ("the material of a thousand uses") were accessible for many different types of application. Although there had been similarly versatile predecessors (for example Bois durci and papier mâché products[2]), it was only with the advent of the new synthetic materials that the aspect of multiple applications gained wide-ranging importance. This aspect also shaped some of the new questions arising in the context of how to design the corresponding mass-produced goods. According to Hans Ulrich Kölsch, the fact that Leo Hendrik Baekelands filed his patent on phenolic resin in the same year that the Deutscher Werkbund was founded, in 1907, was probably just a coincidence, but, on the other hand, symptomatic of an era where people were continually faced by new developments. The honourable and renowned members of the Werkbund, an association of architects, artists and industrialists, had made it their mission to have a reforming and enlightening influence on all fields of the sciences and the arts. As a protest against historicising pomp, art nouveau seemed to have been overcome or was at least in retreat, its tender flowers having only briefly coiled themselves around the long-established machine era. The new

Neben den eigentlichen – später meist hydraulischen –
Pressen, die ihre Vorbilder in den entsprechenden Pressen für die
Kunststoffvorläufer Bois Durci und Papiermaché hatten,[6] waren
noch die beheizbaren, zwei- oder mehrteiligen Edelstahl-Press-
formen notwendig. Das Presspulver wurde in die Matrize (Füll-
form, Unterstempel) gefüllt, diese dann mit der Patrize (Quetsch-
form, Oberstempel) verschlossen und verpresst. Danach wurde
der Gegenstand mittels einer bestimmten Vorrichtung ausgewor-
fen und entgratet.

Die ‚Gute' Pressform

Um nun ein Werkstück prozessgerecht pressen zu können,
mussten Regeln beachtet werden. Unter vielen anderen sind fol-
gende zu nennen:

1) Das Werkstück sollte eine möglichst gleichmäßige
Wanddicke haben. Dazu sind vor allem Verrundungen geeignet.
Konkret heißt dies, dass Ecken und Kanten am besten mit inne-
rer und äußerer Rundung gestaltet werden. Dadurch bleiben die
Wandstärken gleich und die Spannungen im Material werden he-
rabgesetzt.[7] Scharfe Knicke würden demnach die Kerb-Wirkung
begünstigen und erhöhten damit die Bruchgefahr. Somit ergeben
sich bei Abrundungen festere Pressteile.[8]

2) Für eine einfache, schnelle und kostengünstige Entfor-
mung sind Hinterschneidungen zu vermeiden.

3) Um an bestimmten Stellen ein Zusammenziehen des
Materials beim Erkalten außerhalb der Form (Einsinkstellen) zu
kaschieren oder um Materialverstärkung zu erreichen, können
Rippen vorgesehen werden.[9]

Der Matrizen-Form-Stil

Aus Regel Nr. 1 und 2 ergeben sich Gestaltungsmerkmale, für die runde Ecken und gewölbte Kanten charakteristisch
sind. Man kann hier also in der Tat von einem Matrizen-Form-Stil
sprechen. In den USA waren zum Beispiel Walter Dorwin Teague
und Norman Bel Geddes in dieser Hinsicht stilprägend. Teague
formulierte geeignete Ecklösungen (Abb. 1).[10]

Viele Beispiele für eine solche kunststoffgerechte Gestaltung mit Press-Phenolharzen lassen sich aus verschiedenen Ländern und für verschiedene Produkte anführen, wie Kameras

century had presented new challenges: the tre-
mendous increase in population numbers called for
simpler, more radical solutions to social problems.
Transport and production of consumer goods had
to follow in equally radical ways.[3]

The proximity to industrial production enforced a form that had to be in line with the technical requirements and function of a product.[4] This
happened on a large scale with the new synthetic
materials (phenolic resin and urea resin) and the respective processing techniques, i.e. moulding under
high temperatures (phenolic resin at 140-180° C,
urea resin at 140-145° C) and high pressure.[5]

Besides the original – later mostly hydrau-
lic – presses, which were modelled on the presses
for the plastic predecessors, Bois Durci and papier
mâché,[6] heatable stainless steel moulds consisting
of two or more parts were also needed. The press
powder was poured into the matrix (blank mould,
bottom punch), which was then closed and com-
pressed with the male mould (compression mould,
top punch). Eventually, the product was ejected and
de-flashed using special equipment.

The ‘Good' Press Form

In order to mould a component in a pro-
cess-optimised way, a number of criteria had to
be met. Among many others, the following criteria
were important:

1) The component had to have as uniform a
wall thickness as possible. This means that sharp
corners and edges had to be avoided wherever possible by designing corners and edges with inner
and outer curves. This guarantees uniform wall
thickness and lowers the number of stress-points

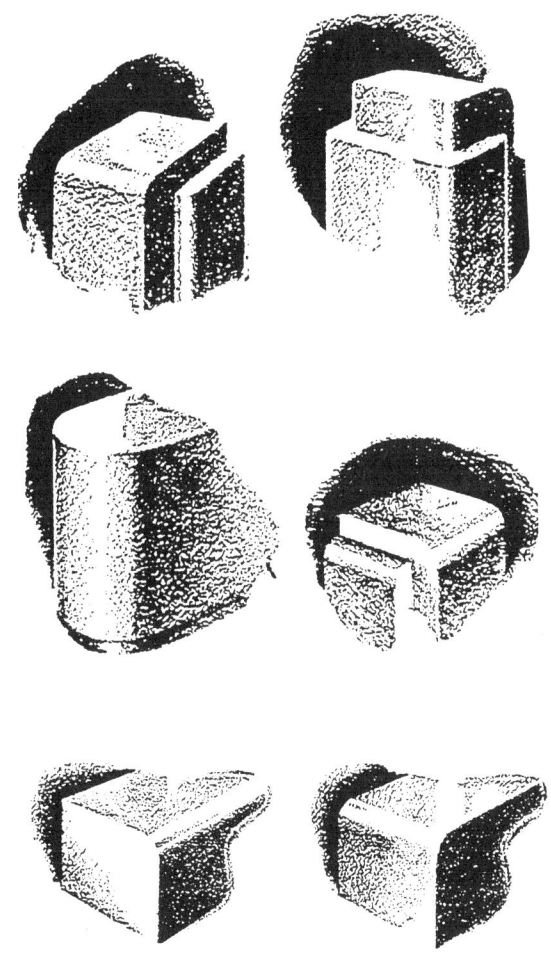

1 / Ecklösungen nach / Solutions for the design of corners by Walter Dorwin Teague, Sammlung / Collection Kölsch.

(Abb. 2 und 3), Leuchten (Abb. 4) und Radios (Kat.Nr. 7 und 8). Die runden Ecken wurden sogar betont, wenn das Phenolharz nicht gepresst, sondern gegossen wurde (Kat.Nr. 28).

Die Übernahme von Gestaltungsregeln für das Formen von Kunststoffpresslingen ging darüber hinaus um vieles weiter. Ohne Pressharz zu verwenden, wurde exakt die gleiche Formensprache auch auf andere Materialien wie Holz oder Metall übertragen, so zum Beispiel beim Radiomodell „Sparton 5518, Selectronne" (Design Walter Dorwin Teague, Kat.Nr. 15). Ebenso wurden Möbel und Architektur in diesem Formenstil gestaltet,

in the material.[7] Sharp bends would lead to stress concentration, thus raising the risk of cracking. Hence curved or rounded designs result in stronger components.[8]

2) Undercuts must be avoided to guarantee easy, time- and cost-efficient ejection.

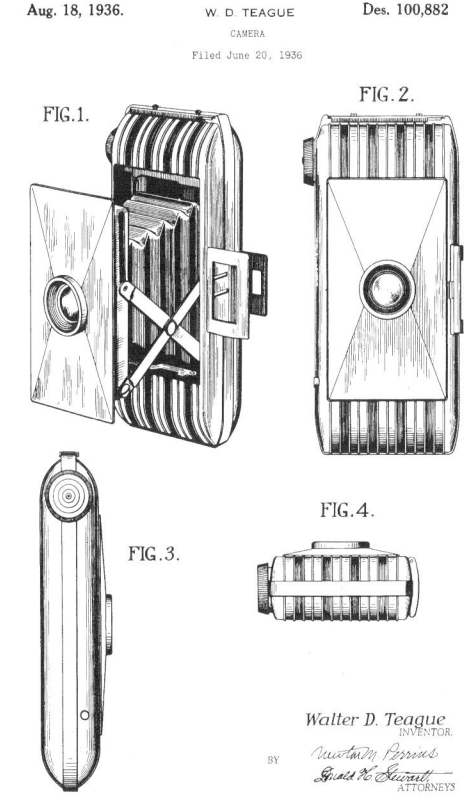

2 / US Patent Nr. D100882S Fotokamera, Walter Dorwin Teague, Eastman Kodak Company, Rochester, New York (US), 1936 © United States Patent and Trademark Office.

3 / Bantam Special, Walter Dorwin Teague, Eastman Kodak Company,
Rochester, New York (US) 1936, Inv.Nr. K 1426 W, MAKK
© Foto / Photo: Sascha Fuis Fotografie, Köln / Cologne.

5 / Shell-Haus, Reichpietschufer 60–62, Berlin (DE),
Emil Fahrenkamp, 1930-32 © Foto / Photo: Beek100.

wie zum Beispiel das „Shell Haus" in Berlin (Abb. 5).[11]

Für diese letzten Beispiele waren die bei Kunstharzen unerlässlichen konstruktiven Merkmale technisch nicht notwendig. Für den Matrizen-Form-Stil kann also mit Fug und Recht gelten: Form follows plastics.

3) Ribs can be used to conceal contraction points in the material (caving) during the cooling process after ejection or to reinforce the material.[9]

The Matrix-Form Style

Criteria 1 and 2 result in designs characterised by rounded corners and bevelled edges. One can, indeed, call this a matrix-form style. In the USA, Walter Dorwin Teague and Norman Bel Geddes, for example, were among those who defined this particular style. Teague developed suitable solutions for the design of corners (fig. 1).[10] There are many ex-

4 / Reklame / Advertisement, Tischleuchte / table lamp,
Christian Dell, Heinrich Römmler AG, Spremberg (DE),
1929/30, Sammlung / Collection Lattermann.

Der Matrizen-Ornament-Stil

Aus Regel Nr. 3 ergeben sich Gestaltungsmerkmale, für die eine sparsame Rippenornamentik charakteristisch ist. Man kann hier von einem Matrizen-Ornament-Stil sprechen. Die Rippen werden nicht mehr nur als ein Versteifungselement und zur Verhinderung von Flusslinien im Material eingesetzt, sondern auch zur Vermeidung großer, ebener Flächen, die nie einfach und kostengünstig zu erreichen sind. Hieraus ergibt sich eine vielfältig-variable, einfache, sparsame, charakteristische Art der Oberflächenverzierung (Abb. 6),[12] die in Bezug auf die Kunstharze zwar technisch nicht mehr zwingend, dafür aber stilbildend war.

6 / Box 14 „Trolix", Agfa-Camera-Werk, München / Munich (DE) 1936-40,
© Foto / Photo: Koppi2.

amples from different countries where suitable designs have been developed for various plastic products made from pressed phenolic resin such as cameras (fig. 2 and 3), lamps (fig. 4) and radios (cat. no. 7 and 8). The rounded corners were even emphasised when the phenolic resin was cast instead of pressed (cat.no. 28).

The design criteria for moulded plastic components were also transferred to other materials. Without using pressed resin, the very same design language was used in products made from wood or metal, for example in the "Sparton 5518, Selectronne" radio (designed by Walter Dorwin Teague, cat.no. 15). Furniture and buildings were also designed in this style, for example the Shell Building in Berlin (fig. 5).[11]

In all the aforementioned designs, the constructional characteristics essential to using synthetic resins were not required on technical grounds. Hence, regarding the matrix-form style, we can rightly state: form follows plastics.

The Matrix-Ornament Style

Criterion 3 results in designs characterised by a considerate use of ornamental ribs. We could call this matrix-ornament style. The ribs are not only used for reinforcement and for avoiding flow lines in the material, but also for avoiding large, even surfaces, which are difficult and costly to implement. Hence the ribs were a versatile, simple, cost-efficient and characteristic ornament (fig. 6)[12], which, although no longer technically mandatory for synthetic resins, defined the matrix-ornament style.

Like the matrix-form style, this ornamental language was also transferred, in many differ-

Analog zum Matrizen-Form-Stil – wurde die Sprache dieser Ornamentik auch auf Objekte außerhalb des Kunststoffbereichs in vielfältiger Art übertragen. Neben Pressglas war dies ebenso für Metallobjekte und in der Architektur der Fall.[13] Für den Matrizen-Ornamentik-Stil kann also ebenso mit Fug und Recht gelten: Ornament follows plastics.

7 / Schale und zwei Vasen / Bowl and two vase, Frankreich / France um / about 1925, Sammlung / Collection Kölsch © Foto / Photo: Atelier Müller, Essen.

Schluss

Die Formen- und die Ornament-Spielarten des Matrizen-Stils sind einfach und klar, aber weicher und oberflächengestalteter als die geometrisch-strengeren Arbeiten des Bauhauses. Allerdings hat sich das Bauhaus nicht mit der Gestaltung von Kunststoffen beschäftigt.[14] Einen größeren stilistischen Abstand zum Matrizen-Stil gibt es dagegen zu den eleganten oder expressiven Varianten des Art Déco. Dies wird beispielsweise in Frankreich an der ausdrucksstarken Gestaltung von der am Kubismus orientierten Formensprache (Abb. 7), dem ‚Zacken-Stil‘, und dem Radiomodell „66 Skyscraper" der Firma Air-King Products Company im ‚Stufen-Stil‘ deutlich (Kat.Nr. 9).

ent ways, to non-plastic products for example to pressed glass and metal and to architecture.[13] Therefore, regarding the matrix-ornament style, we can again state: ornament follows plastics.

Conclusion

The forms and ornaments of the matrix style are related to the New Objectivity expressed in many geometric-linear Bauhaus pieces.[14] The elegant or expressive designs of art déco, on the other hand, are further removed from the matrix style. Examples for this can be found in French design, in expressive pieces using a design language based on cubism (fig. 7), in the zig-zag style and in the stepped design of the "66 Skyscraper" radio from Air King Products Company (cat.no. 9).

Later forms of the matrix style transitioned in the 1940s and '50s into streamline modernism, like, for example, the "Bolide" table lamp based on a 1946 design by André Mounique (fig. 8).[15]

The design of products made from pressed phenolic resin in the years before the Second World War has had a great influence on the development of product design, right up to the present with its much greater variety of products.[16] Furthermore, the pioneering work of these early plastic designers together with the corresponding technical requirements also defined the design of other materials.

1   See Kölsch, 1993: 81-93.

2   See Lallemand, 1999.

3   See Kölsch, 1993: 81-93.

4   See Crespy/Bozonnet/Meier, 2008: 3368-3374.

5   See Mehdorn, 1939: 91-92 and Bandenburger 1938: 103-124.

6   See Chevalier, 1862: 69-71 and Vermosen [n.year 2008]: 41.

7   Bonten, 2003: 122.

Hinsichtlich des Matrizen-Form-Stils sind dessen spätere Formen in den 1940er und 50er Jahren unmittelbar in die Stromlinien-Moderne übergegangen, wie die klappbare Tischlampe „Bolide" nach einem Entwurf von André Mounique von 1946 verdeutlicht (Abb. 8).[15]

Die Gestaltung von Pressharz-Objekten durch frühe Kunststoffdesigner in den Jahren vor dem Zweiten Weltkrieg beeinflusste die von da an fortlaufende und bis heute gewaltig angewachsene Entwicklung des Produktdesigns.[16] Darüber hinaus war die Pionierarbeit dieses Kunststoffdesigns im Zusammenspiel mit den jeweiligen technischen Erfordernissen maßgebend für die Gestaltung anderer Materialien.

1   Vgl. Kölsch, 1993, S. 81-93.

2   Vgl. Lallemand, 1999.

3   Vgl. Kölsch, 1993, S. 81-93.

4   Vgl. Crespy/Bozonnet/Meier, 2008, S. 3368-3374.

5   Vgl. Mehdorn, 1939, S. 91-92 und vgl. Bandenburger 1938, S. 103-124.

6   Vgl. Chevalier, 1862, S. 69-71 und vgl. Vermosen [o.J. 2008], S. 41.

7   Bonten, 2003, S. 122.

8   Nitsche/Esch, 1940.

9   Boten, 2003, S 123.

10 Vgl. Kölsch, 1993, S. 81-93.

11 Shell Haus, Berlin, Emil Fahrenkamp, 1930-32 und die Römerstadt, Frankfurt a. M., Ernst May, 1927-29.

12 Vgl. DuBois, 1943, S. 256.

13 Vgl. Mundt, 1991, S. 94 und Manske/Scholz, 1987, S. 126.

14 Lattermann, 2003.

15 André Mounique: Retractable Office Lamp, US-Patent 2543926 A 1946/51. Vgl. Hornik, 2011.

16 Lattermann 2006.

8   Nitsche/Esch, 1940.

9   Boten, 2003: 123.

10 See Kölsch, 1993: 81-93.

11 Shell Haus, Berlin, Emil Fahrenkamp, 1930-32 and Römerstadt district, Frankfurt am Main, Ernst May, 1927-29.

12 See Dubois, 1943: 256.

13 See Mundt, 1991: 94 and Manske/Scholz, 1987: 126.

14 Lattermann, 2003.

15 André Mounique: Retractable Office Lamp, US-Patent 2543926 A 1946/51. See Hornik, 2011.

16 Lattermann 2006.

8 / Bolide (Lucidus Bloc), André Mounique, Jumo Breveté, Paris (FR) 1946, Inv.Nr. K 729, MAKK
© Foto / Photo: Rheinisches Bildarchiv Köln / Cologne, Marion Mennicken.

Die Suche nach der Guten Form.
Radio-Design und die Firma Braun in
den 1950er Jahren

In Search of 'Good Form'.
Radio Design and the Braun Company in the 1950s

Elina Knorpp

Nach dem Zweiten Weltkrieg wurde das Radio zu einem festen Bestandteil von Küchen und Wohnzimmereinrichtungen. Radios, später Fernseher, Staubsauger, Kühlschränke und Tonbandgeräte wurden zu begehrten Gegenständen der modernen Wohnkultur und fanden sogar Eingang in die Bildende Kunst, so zum Beispiel in den Werken des Pop Art-Künstlers Tom Wesselmann. Man experimentierte mit der Form dieses modernen Mediums und ging sogar so weit, Jugendfahrräder und Kühlschränke mit integrierten Radios zu produzieren.

Die ökonomische Situation im Nachkriegsdeutschland stellte sich anders dar als etwa in den USA oder in England. Doch bereits innerhalb eines Jahrzehnts zeichneten sich in Deutschland zunehmender Fortschritt, Wohlstand und damit ein Wirtschaftswachstum ab, das als ‚Wirtschaftswunder' bezeichnet wird. Gleichzeitig fand nach Friedrich Lenger die sogenannte „Verhäuslichung" statt; man verbrachte die Freizeit und Abende gerne zu Hause, durch Radio und später Fernsehen war dabei der private „Konsum von Kultur- und Unterhaltungsangeboten [möglich], die zuvor das Verlassen der Wohnung vorausgesetzt hatten"[1]. Die Einrichtung der Wohnzimmer folgte diesem Trend und machte Platz für die neuen technischen Geräte. In einem Ratgeber aus dem Jahre 1960 „Unsere Wohnung. Einrichten und Gestalten" heißt es: „Radio und Plattenspieler haben längst ihren festen Platz im Wohnraum: gut zu bedienen, also entweder niedrig in Nähe des Sessels, oder aber wenn man das Gerät entfernter haben will auch in einer Regalzone so hoch untergebracht, dass man im Stehen auflegt und einstellt. Gern lässt man diese Musikgeräte in der reinen Zweckform gut proportionierter Kästen in den Regalfächern von Bücherwänden verschwinden."[2]

Doch bei zahlreichen Geräten handelt es sich zu Beginn der 1950er Jahre nicht immer um die beschriebenen „gut proportionierten Kästen" oder schlichte Möbel. Man begegnet dagegen häufig dem Phänomen des sogenannten ‚Gelsenkirchener Barocks', wobei es sich um eine abwertende Bezeichnung für industriell hergestellte Möbel der 1930er bis 1950er Jahre handelt, die heute als Beispiel des schlechten Geschmacks gelten.[3] Nachgeahmt wurden dabei eklektisch-historisierende Möbel des 19. Jahrhunderts – massiv, oft mit stark betonten Wülsten und gewellten Türen. Die Gestalt von Möbeln und Radiogeräten beeinflusste sich dabei gegenseitig und so wurden passend

After the Second World War, the radio became a permanent feature in kitchens and living rooms. Radios – and later TV sets, vacuum cleaners, refrigerators and tape recorders – became sought-after products of modern domestic culture and they even featured in the visual arts, for example in pop art pieces by Tom Wesselmann. Many experiments were carried out with the form of the modern medium of radio, even going as far as to produce bicycles for young consumers and refrigerators with integrated radios.

The economic situation in postwar Germany was different to that in the USA or in England, for example. However, within the space of one decade, Germany was experiencing increasing progress and prosperity and an economic growth known as the Wirtschaftswunder. According to Friedrich Lenger, this development was paralleled by a trend towards 'cocooning': people took to spending their free time and evenings at home. Through radio and, later, television it was no longer necessary to leave one's home if one wanted to enjoy culture and entertainment programmes.[1] The interiors of living rooms followed this trend by accommodating the new devices. In a guidebook from 1960 with the title Our Home. Furnishing and Design, it says: "Radios and turntables have long become permanent features in the living room: easy to use, either placed at a low level near the armchair or – if one prefers to have the devices further away – positioned on a high shelf so that one can put records on the turntable and switch on the device while standing up. With their practical forms of well-proportioned chests one likes to hide these music systems in the shelves of wall-sized bookcases."[2]

zur Einrichtung massive Radios, oft aus Nussbaumholz oder -furnier, mit messing- oder goldfarbenen Dekorelementen und Leistenprofilen versehen. Außerdem wurden Radioapparate in Schränke eingebaut und folglich als ‚Phonoschränke' bezeichnet. So verschwand das technische Gerät im Inneren und konnte das Interieur nicht ‚stören'.

Nach dem Krieg versuchte die Möbelindustrie zwar noch an die Gestaltung der 1930er Jahre anzuknüpfen. Es bedurfte jedoch neuer Formen und leichter, funktionaler Möbel, da im Krieg über fünf Millionen Wohnungen zerstört worden waren und sich die Wohnsituation im Allgemeinen sehr beengt darstellte. Die Frage nach einer guten Gestaltung und einer guten Form beschäftigte vor allem professionelle Kreise. In Deutschland war bereits 1907 eine Institution eingerichtet worden, die sich der Aufgabe stellte, die ‚Gute Form' zu definieren und sie dem Konsumenten durch Publikationen und Ausstellungen näher zu bringen: der Deutsche Werkbund.[4] Auch der 1951 in Darmstadt gegründete Rat für Formgebung förderte das deutsche Design. Beide Institutionen traten für Funktionalität und Materialgerechtigkeit, für die sogenannte ‚Gute Form'[5] ein und lehnten historisierende und organische Formen sowie alles rein Schmückende ab.

Im Bereich der Möbelgestaltung kamen neue Impulse sowohl aus Skandinavien schlichte Naturholzmöbel, besonders aus Teakholz als auch aus den USA. Hier sind insbesondere die Unternehmen Knoll International[6] und Herman Miller Furniture Company zu nennen, die mit den Designern Charles und Ray Eames, Eero Saarinen und Harry Bertoia zusammenarbeiteten sowie die Möbel von Marcel Breuer und Ludwig Mies van der Rohe wieder auflegten.

Die Firma Braun

Im Kontext dieses neuen Einrichtungsstils sollten auch die Radio- und Phonogeräte der Firma Braun gestaltet werden. Nachdem der Gründer der Firma, Max Braun, 1951 plötzlich verstorben war, übernahmen seine Söhne Artur und Erwin im Alter von erst 26 beziehungsweise 30 Jahren die Leitung. Die Firma „Max Braun. Maschinen- und Apparatebau" (gegr. 1921), die spätere Braun AG, erhielt ab 1932 die Bauerlaubnisgenehmigung für Röhrengeräte und begann als erstes Unternehmen in Europa mit

However, at the beginning of the 1950s, many of these devices did not come in the form of "well-proportioned chests" or simply-styled furniture. Instead, one often encountered what was called 'Gelsenkirchener Barock', a pejorative term for mass-produced pieces of furniture from the period between the 1930s and the 1950s, which are today regarded as examples of bad taste:[3] remakes of 19th century furniture featuring an eclectic array of historical references – massive, heavy pieces, often heavily sculpted and with undulating doors. As the shape of furniture and radio devices mutually influenced each other, bulky radios, often made from solid walnut or walnut veneer, were decorated with brass or gold-coloured elements and bezels to match the furniture. Furthermore, radio devices were integrated into cabinets, which were consequently referred to as 'phono cabinets': the technical device disappeared inside the cabinet so as not to 'disturb' the interior of the living room.

After the war, the furniture industry still clung to the design style of the 1930s. What was needed, however, were new forms and unobtrusive, functional furniture, as five million homes had been destroyed, resulting in rather cramped and spatially restricted living conditions. The question of good design and good form was mainly discussed in the design industry. As early as 1907, a dedicated institution had been established in Germany whose task it was to define 'good form' and to familiarise consumers with good form through publications and exhibitions: the Deutscher Werkbund.[4] Established in 1951, the Rat für Formgebung (German Design Council) also promoted German design. Both institutions championed functionality and an honest use of materials. They supported 'good form'[5] and rejected historicising and organic forms as well as purely decorative elements.

der Produktion einer Kombination von Plattenspieler und Radio dem „Phono-Super Cosmophon 777". Für 1951, dem Zeitpunkt des Generationswechsels in der Firmenleitung, charakterisiert Hans Wichmann das Unternehmen wie folgt: „Alles, was nach außen wirkte, die Werbung, die Erscheinung der Geräte […] war konventionell, durchaus anständig, solide, weit entfernt von Besonderheit."[7] Die Brüder wollten das Unternehmen grundlegend verändern. Während der Ingenieur Artur Braun die Aufgaben der technischen Entwicklung übernahm, beschäftigte sich Erwin Braun sowohl mit der Konzeption neuer Produkte, als auch mit dem einheitlichen Unternehmensbild.[8]

Eine Reihe von Anregungen beeinflusste dabei die Suche nach der neuen Form. Zum einen gehörte dazu die Zusammenarbeit mit Fritz Eichler, einem Kriegskameraden Erwin Brauns, der als Filmregisseur 1953 für die Produktion der Werbefilme engagiert wurde und dem Unternehmen beratend zur Seite stand.[9] Artur Braun schrieb: „[Fritz Eichler und Erwin Braun] versuchten einen gemeinsamen Nenner zu finden, eine gemeinsame Aussage für unser doch so heterogenes Produktionsprogramm."[10]

Zu weiteren Inspirationsquellen gehörte Erwin Braun zufolge ein Vortrag des ehemaligen Bauhausschülers Wilhelm Wagenfeld 1954 in Darmstadt, der über die Qualitäten eines Industrieerzeugnisses sprach. Dieses müsse „sein eigenes Wesen" haben und „von individuellen Einströmungen […] völlig geläutert" werden.[11] Erwin Braun bot Wagenfeld daraufhin eine Zusammenarbeit in der Produktgestaltung an. Einen wichtigen Entwicklungsschritt in der Design-Geschichte Brauns stellt darüberhinaus die Kooperation mit der Hochschule für Gestaltung in Ulm (HfG) dar, die 1954 ihren Anfang nahm.

Die Zusammenarbeit mit der HfG Ulm

Im Jahr 1955 hielt Walter Gropius bei der offiziellen Einweihung des Gebäudekomplexes der zwei Jahre zuvor gegründeten Hochschule für Gestaltung Ulm eine Festrede.[12] Die Anwesenheit des Bauhaus-Gründers hatte dabei einen symbolischen Charakter, basierte doch die Gründung der Schule auf der Idee der Anknüpfung an das Bauhaus und seiner Prinzipien. Der Mitbegründer der Hochschule, Otl Aicher, entwickelte das soge-

In furniture design, new impulses came from both Sweden (simple wooden furniture, mainly made from teak) and the USA. In this context, we would have to name in particular the companies Knoll International[6] and the Herman Miller Furniture Company, which collaborated with designers Charles and Ray Eames, Eero Saarinen and Harry Bertoia and also relaunched furniture by Marcel Breuer and Ludwig Mies van der Rohe.

The Braun Company

The radio and phono devices from Braun would soon also be designed in the context of the new interior design style. After the company's founder, Max Braun, unexpectedly passed away in 1951, his sons Artur and Erwin took over, aged 26 and 30 respectively. The "Max Braun. Maschinen- und Apparatebau" company (establ.1921), which would later become Braun AG, was, in 1931, granted permission to built valve radios and became the first company in Europe to start producing a combination of turntable and radio: the "Phono-Super Cosmophon 777". Referring to the year 1951, when Max Braun's sons took over the company management, Hans Wichmann describes the company thus: "Everything defining the company's external image – advertising, the products' looks […] – was conventional, thoroughly decent, sound, far removed from being special."[7] The brothers wanted to radically change the company. While engineer Artur Braun took over technical development, Erwin Braun focused on developing new product concepts, as well as a coherent company image.[8]

A number of stimuli influenced the search for good form. For example, the collaboration with Fritz Eichler, a wartime comrade of Erwin Braun's, who, being a film director, was commissioned in 1953 to

nannte ‚Ulmer Modell', bei dem neben Lehre und Forschung auch die Produktentwicklung berücksichtigt wurde – „das Hervorbringen praxisfähiger und serienreifer Prototypen für Auftraggeber"[13]. Fortgeschrittene Studenten arbeiteten dabei unter Leitung der Dozenten in verschiedenen Forschungsgruppen.[14]

Die Entwicklung des Produktdesigns für Braun übernahm zusammen mit seinen Studenten und Mitarbeitern Hans Gugelot, holländischer Architekt und Designer sowie Dozent für Produktgestaltung an der HfG Ulm. Das funktionalistische Ideal der Ulmer Hochschule, das mit den Merkmalen technisch, einfach und zurückhaltend beschrieben werden kann, passte ideal zur Intention der Braun-Leitung: „Unsere elektrischen Geräte sollen unaufdringliche, stille Helfer und Diener sein."[15]

Radios von Braun

Radioapparate der frühen 1950er Jahre sind entweder aus Kunststoff hergestellt und weisen eine glänzende Oberfläche und abgerundete Formen mit verspielten, manchmal goldfarbenen Elementen auf (Kat.Nr. 79) oder sie besitzen ein Gehäuse aus lackiertem Holz und erinnern somit an die Geräte der 1930er Jahre (Kat.Nr. 100). Wie Fritz Eichler formulierte, „schrien" die Radiogeräte geradezu danach, „von ihrer goldprotzenden Verlogenheit befreit zu werden"[16]. Erwin Braun erklärte rückblickend: „Bei Radiogeräten fangen wir am Nullpunkt an. Wir kannten keine Muster, wie gut gestaltete Radios aussehen sollen."[17]

Das erste von Artur Braun und Fritz Eichler für Braun neu entworfene Radio ist der „Kleinsuper SK 1" – ein schlichtes Gerät mit einem Kunststoffrahmen und einer Lochplatte als Front (Kat. Nr. 101). Direkt auf die Lochplatte sind zwei Bedienungsknöpfe und eine runde Skala angebracht. Die Front mit den runden Elementen ist klar geordnet. Das Aussehen leitet sich dabei von der Funktion ab und hat kein dekoratives Beiwerk.

Doch obwohl man Erwin Braun zufolge mit der Gestaltung bei Null anfing, kann man sicher von gewissen Vorbildern sprechen. Auffällig ist zum Beispiel eine Ähnlichkeit mit dem amerikanischen Transistorradio „Regency TR-1", das ein Jahr früher, 1954, als erstes kommerzielles Transistorradio in den Verkauf ging (Abb. 1).[18] Es ist ebenfalls auf die Formen des Rechtecks und des Kreises reduziert, allerdings erscheint das Gerät durch

produce the company's commercials and to provide consulting.[9] Artur Braun is quoted as stating: "[Fritz Eichler and Erwin Braun] tried to find a common denominator, a coherent message for our rather heterogeneous production programme."[10]

According to Erwin Braun, another source of inspiration was a talk given in 1954 in Darmstadt by former Bauhaus student Wilhelm Wagenfeld. Talking about the qualities of a mass-produced product, Wagenfeld stated that such a product would need to have "its own character" and had to be "entirely liberated of individual influences."[11] Following this talk, Erwin Braun offered Wagenfeld the opportunity to collaborate in the design of Braun products. Another important step in the development of Braun's design history was the collaboration with the Ulm School of Design (HfG), which started in 1954.

Collaborating with the HfG Ulm

In 1955 Walter Gropius gave a speech at the official inauguration of the buildings that would house the Ulm School of Design, which had been founded two years earlier.[12] The presence of the Bauhaus founder had a symbolic character as the school's founding was based on the idea of connecting to the Bauhaus and its principles. A co-founder of the Ulm School, Otl Aicher developed what is known as the 'Ulm model', which, besides teaching and research, also focused on product development: "the creation of usable, feasible and production-ready prototypes for clients."[13] To this end, advanced students would work in different research groups under the supervision of teachers.[14]

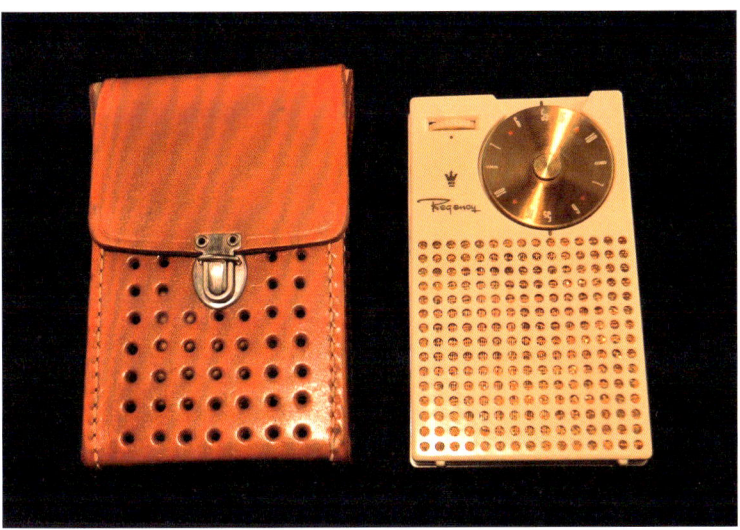

1 / Regency TR-1, Industrial Development Engineering Associates (I.D.E.A.), Indianapolis, Indiana (US), 1954 © Foto / Photo: Joe Haupt.

die metallische Senderwahlscheibe nicht so puristisch wie das von Braun. Im Jahr 1957 gewann die Firma Braun auf der XI. Design-Triennale in Mailand den Grand Prix und das weiterent-wickelte Model „Kleinsuper SK 2" durfte in der Vitrine der Nationen ausgestellt werden, in der besonders gelungene Produkte präsentiert wurden. Dieses Modell sollte ein Teil des zukünftigen Systems der miteinander kompatiblen Radio- und Phono-Geräte von Braun werden.[19] Beim „Kleinsuper SK 2" war es zum Beispiel möglich, einen Plattenspieler anzuschließen und ihn als Lautsprecher zu verwenden.[20]

Formal gibt es auch eine Verbindung zwischen dem „Kleinsuper SK 1" und dem Transistorradio „Exporter 2", das 1956 durch die HfG Ulm auf der Basis des älteren Models „Exporter 1" (1954) neu gestaltet wurde (Abb. 2). Beim „Exporter 1" handelt es sich um ein tragbares Kofferradio. Die Front gliedert sich in einen Lautsprecher mit schlitzartigen Öffnungen und eine Wahlscheibe sowie einen Lautsprecherregler. Der „Exporter 2" hat die gleiche Form, die Gestaltung der Front ist auf der Basis schon vorhandener Elemente noch klarer und reduzierter. Die Oberfläche ist glatt und die Lautsprecheröffnung auf ein Rechteck reduziert. Die Zahlen, die schon beim „Exporter 1" auf der Front außerhalb der Scheibe angebracht waren, sind kleiner und in einer klaren

Together with his students and colleagues, Hans Gugelot, a Dutch architect and designer and professor of product design at HfG Ulm, took over the product design development for Braun. The functionalist ideal of the Ulm School, which can be described with the attributes 'technical', 'simple' and 'unobtrusive', perfectly matched the intentions of the Braun management: "We want our electrical devices to be unobtrusive, silent helpers and servants."[15]

Braun Radios

Radios from the early 1950s are either made of plastic, featuring high-gloss surfaces and rounded forms with playful, sometimes gold-coloured elements (cat.no. 79) or they have a wooden casing with a lacquer finish, thus being reminiscent of devices from the 1930s (cat.no. 100). As Fritz Eichler stated, the radios were virtually 'pleading' to be liberated from their ostentatious gold décor hypocrisy."[16] Erwin Braun retrospectively explained: "With radio devices, we started with a clean slate. We did not know any patterns telling us what well-designed radios should look like."[17]

The first new radio designed by Artur Braun and Fritz Eichler was the "Kleinsuper SK 1", a simple device featuring a plastic casing and a perforated front panel (cat.no. 101). Two controls and a circular dial are mounted directly on the front panel – a clear and simple arrangement. The radio's looks are derived from its function, without any decorative adornment.

Although, according to Erwin Braun, the design started with "a clean slate", it seems fair to draw references to certain predecessors. For example, there is a striking similarity to the US "Regency TR-1" transistor radio, which had been launched

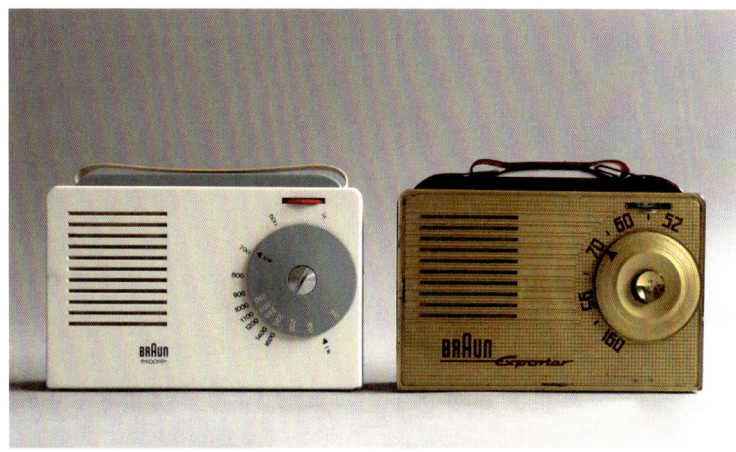

Typographie gestaltet.[21] Der „Exporter 2" ist in Weiß in Kombination mit Grau-Blau oder Orange ausgeführt, mit einem Akzent auf dem roten Lautsprecherregler. Die zur Auswahl stehende Farbpalette der Produkte bei Braun stellte überhaupt eine Neuheit dar. Dezente Farben wie helles Grau und Weiß kombiniert mit modischen Tönen, wie Hellgrün, Hellblau oder Orange, erschienen modern und passten zu den neuen Einrichtungen. Später wurde die Kombination von Metall-Grau und Schwarz verwendet, die sich für HiFi-Anlagen unter anderem bis heute durchgesetzt hat.

Die weiteren Modelle der portablen Radios und Tischgeräte von Braun, wie der „T 2" (1958) oder der „RT 20" (1961) von Dieter Rams (Kat.Nr. 99 und 111), der ab 1955 für Braun arbeitete und 1961 Chefdesigner wurde, griffen die Elemente und Struktur der genannten Modelle auf und sorgten für eine Wiedererkennbarkeit der Braun-Produkte.

Das System von Braun beinhaltete nicht nur eine technische Kompatibilität, durch die die Geräte miteinander verbunden werden konnten, sondern auch die Möglichkeit, diese zu kombinieren, sie nebeneinander oder aufeinander zu stellen, so zum Beispiel das Radio „G 11", der Plattenspieler „G 12" und das Fernsehgerät „FS-G".[22] Neben dem „G 11" entwarf Hans Gugelot 1955 den „Tischsuper TSG" und die Serie der Kompaktgeräte „PKG", genannt „Langer Heinrich", die sich durch Holzgehäuse, grafisch

a year earlier, in 1954, as the first commercial transistor radio (fig. 1).[18] The "Regency TR-1" is also reduced to the shapes of rectangle and circle, but with its metal-finish dial, its looks are not as purist as the Braun model's. In 1957, Braun won the Grand Prix at the XI. Design Triennial in Milan and the "SK 1"'s successor, the "SK 2", was allowed into the Showcase of the Nations, where only particularly well-designed products were exhibited. The "SK 2" would become a part of the future system of compatible radio and phono devices from Braun.[19] With the "SK 2" it was, for example, possible to connect a turntable and use the radio as a speaker.[20]

In terms of form, there is also a connection between the "SK 1" and the "Exporter 2" transistor radio, which, based on its predecessor "Exporter 1" (1954), was redesigned at the HfG Ulm in 1956 (fig. 2). The "Exporter 1" was a portable radio. Its front features a speaker with slits, a dial and a volume control. The "Exporter 2" has the same form while, based on existing elements, the front design is even clearer and more reduced. The digits, which were already placed not on, but around, the dial in the "Exporter 1", are smaller and designed with clear typography.[21] The "Exporter 2" was available in white with either grey-blue or orange, with its volume control accentuated in red. The choice of available colour schemes for Braun products was a total novelty, with subtle colours, such as light grey and white combined with fashionable shades, such as light green, light blue or orange, giving the products a modern look while also matching the new interiors. Later, a combination of metallic grey and black was used, which, in the design of HiFi systems, has prevailed to the present day.

gestaltete Skalen sowie eine klare Anordnung der Bedienelemente auszeichnen (Kat.Nr. 87 und 88). Bei den „PKG"-Geräten handelt es sich um Musikschränke, die auf einem H-förmigen Gestell angebracht sind, in dem auch eine Ablage für Schallplatten und Zeitschriften vorgesehen ist. Das Gerät kann aber herausgenommen und zum Beispiel in einem Regal aufgestellt werden.

Die neuen Braun-Radios, die innerhalb von acht Monaten gestaltet wurden, präsentierte man auf der Deutschen Rundfunk-, Fernseh- und Phono-Ausstellung 1955 in Düsseldorf. Der Stand wurde vom HfG-Professor Otl Aicher und seinen Mitarbeitern konzipiert (Abb. 3). Hans Gugelot erinnerte sich an die Wirkung des Messestandes: „dieses hell beleuchtete objekt inmitten der mit fontänen und guirlanden geschmückten stände der konkurrenz löste einen wahren schock aus [sic]."[23] In der Presse wurde sehr viel über den Braun-Auftritt auf der Messe und seine sensationelle Wirkung berichtet. Doch die Reaktionen waren nicht nur positiv: Max Grundig soll zu den Braun-Söhnen gesagt haben: „Was macht ihr da mit dem Erbe eures Vaters?"[24]

3 / Braun-Messestand auf der Funkausstellung / Stand at the radio show, Düsseldorf (DE), 1955 © Braun Sammlung.

Further models of portable and transistor radios from Braun picked up the elements and structure used in the aforementioned designs, thus making Braun products easily recognisable. These later models include the "T 2" (1960) and "T 41" (1962) designed by Dieter Rams (cat.no. 99 and 111), who joined Braun in 1955 and became the company's head designer in 1961.

The Braun system not only ensured technical compatibility that allowed different devices to be connected to each other, but also allowed the combining of components by putting them next to, or on top of, each other, for example the "G 11" radio, the "G 12" turntable and the "FS-G" TV set.[22] Besides the "G 11", Hans Gugelot also designed the 1955 "Tischsuper TSG" and the "PKG" range of compact devices, nicknamed 'Langer Heinrich', which are characterised by their wooden casings, dials with clear graphics and a clear layout of control elements (cat.no. 87 and 88). The "PKG" devices are radiograms mounted on an H-shaped base that also includes a shelf for storing records and magazines. The device can also be taken out of the cabinet and, for example, put on a shelf.

Designed within the space of eight months, the new Braun radios were presented at the 1955 Deutsche Rundfunk-, Fernseh- und Phono-Ausstellung trade exhibition in Düsseldorf. The Braun booth was designed by HfG professor Otl Aicher and his colleagues (fig. 3). Describing the effect of the exhibition design, Hans Gugelot states: "Among the booths of competitors, which featured water fountains and garlands, this brightly lit object had a truly shocking effect."[23] There was a flurry of press reports about the Braun exhibition at the trade fair and its sensational effect. But not all reactions

Die neue Gestaltung wurde aber auch gewürdigt, so liest man in einer Besprechung aus dem Jahr 1957: „An ihnen [den Radios] ist nichts Überflüssiges, nichts Störendes. Helles Holz in Verbindung mit zeitlos elegant geformten Kunststoffteilen, Drehknöpfe, die auch wirklich aussehen wie Drehknöpfe und nicht wie gefrorene Puderquasten mit Goldlack."[25]

Als Titelbild des Produkt-Katalogs „Radio- und Fernsehgeräte im Stil unserer Zeit" (1956) wurden Braun-Geräte in der modernen Einrichtung mit den Möbeln der Knoll International von den Designern Harry Bertoia, Florence Knoll, Herbert Hirche, Ferrari-Hardoy und anderen aufgenommen.[26] Herbert Hirche, dessen Möbel bereits im Vorfeld für die Präsentation von Braun-Radiogeräten ausgewählt wurden, bekam 1955 den Gestaltungsauftrag von in dieser Zeit beliebten Musiktruhen. Seine Musiktruhen der Serie „HM" wurden in Teak- oder Nussbaumholz ausgeführt und passten zu modernen Möbeln im skandinavischen Stil. Die getrennten Lautsprechergruppen sorgten für Stereo-Wirkung (Kat.Nr. 102).

Schneewittchen-Sarg

Das mit Abstand bekannteste und vielbesprochene Braun-Radio ist der „Phonosuper SK 4", im Volksmund als „Schneewittchen-Sarg" bezeichnet (Kat.Nr. 98). Er fand Eingang in zahlreiche Designsammlungen. Der „Phonosuper SK 4" wurde in Teamarbeit von Hans Gugelot, Dieter Rams und Herbert Lindinger 1956 geschaffen, wobei Wilhelm Wagenfeld den Plattenspieler entwickelte. Es handelt sich um eine Radio-Plattenspieler-Kombination mit einem Lautsprecher (nach Vorlage eines Braun-Radios aus den 1930er Jahren). Der „Phonosuper SK 4" hat eine schlichte Kastenform, bei der ein weißes U-förmig gebogenes Blechgehäuse zwischen zwei Wangen aus poliertem Buchenholz gespannt ist.[27] Eine Neuheit bildete der Plexiglasdeckel, der anfänglich aus Metall entworfen wurde, aber aufgrund der Eigenvibration verworfen wurde. Der transparente Deckel und die rechteckige Kistenform gaben dem Gerät seinen aus der Märchenwelt entlehnten Spitznamen. Die Vorder- und Rückseite ist durch die Lautsprecheröffnungen – Rechteck und Quadrat – im Verhältnis 2/3 zu 1/3 gegliedert. Neben den Bedienelementen in L-Form befindet sich das rechteckige Feld mit dem charakteristischen runden Plattenteller, das auch in weiteren Modellen aufgegriffen wurde. Die Kombination aus Grau und Weiß er-

were positive. Max Grundig is cited as having said to the Braun sons: "What on earth are you doing with your father's heritage?"[24]

However, the new design was also met with critical approval. A 1957 review for example reads: "There is nothing superfluous or disturbing in these radios. Light-coloured wood combined with timelessly and elegantly shaped plastic parts, control dials that really look like control dials and not like frozen powder puffs covered in gold-coloured lacquer."[25]

The cover of the "Radio- und Fernsehgeräte im Stil unserer Zeit" product catalogue (1956) featured Braun devices in modern interiors with furniture from Knoll International designed by Harry Bertoia, Florence Knoll, Herbert Hirche, Ferrari-Hardoy and others.[26] Herbert Hirche, whose furniture designs had already been used earlier for the presentation of Braun radios, was commissioned in 1955 to design the then-popular audio furniture. His "HM" range of audio furniture was produced in teak or walnut, matching modern Scandinavian-style furniture. Separated loudspeaker units created the stereo effect (cat.no. 102).

Snow White's Coffin

The "Phonosuper SK 4", more commonly known as 'Snow White's Coffin', is by far the most well-known and most discussed Braun radio (cat.no. 98), included in many design collections. The " Phonosuper SK 4" was created in 1956 as a collaborative project by Hans Gugelot, Dieter Rams and Herbert Lindinger, with Wilhelm Wagenfeld designing the turntable. The device is a radio and turntable combination with speaker (based on a Braun radio from the 1930s). The "Phonosuper SK 4" features a

scheint hier modern und elegant. Auch das Kombigerät „Atelier 1"
wurde in Anlehnung daran konzipiert (Kat.Nr. 97). Die Bedienung
wird hier wieder auf der vorderen Seite des Geräts platziert und
der Lautsprecher „L 1" ist nicht mehr integriert, sondern
als separate Einheit gestaltet. Die Geräte ergänzen sich in Material
und Gestaltung. Gerd Selle stellt fest, dass der „Schneewittchensarg" als Nachweis des guten Wohngeschmacks galt. „Außer
bei einer schmalen Schicht ästhetisch Gebildeter, die sich an
Werkbundschlichtheit und Bauhausform orientieren konnte,
war dieses Gerät ein mehr oder weniger exotisches Objekt. [...]
seine Strenge [war] eine Zumutung für den berühmten ‚Otto
Normalverbraucher'."[28]

4 / iPod der 5. Generation / 5th generation iPod, Jonathan Ive, Apple,
Cupertino, Kalifornien / California (US) 2005 © Foto / Photo: Ilmari Karonen.

simple chest form, with a white, U-shaped sheet
steel casing braced between two cheeks of polished
beechwood.[27] Originally designed in metal, but discarded due to vibration, the acrylic glass cover was a
novelty feature. The device owes its fairy-tale nickname to its clear cover and rectangular box shape.
Front and back are structured by the speaker openings – rectangle and square – in a 2:3 and 1:3 ratio.
Next to the L-shaped control elements is the rectangular surface with the round turntable, which was
also used in other models. The combination of grey
and white looks modern and elegant. The "Atelier 1"
combination device was also based on this design
(cat.no. 97). In the "Atelier 1", the controls are again
positioned at the front while the "L 1" speaker is not
integrated, but designed as a separate unit. In terms
of material and design, the devices complement
each other. Gert Selle states that the 'Snow White's
Coffin' was regarded as evidence of good taste in
interior design: "Apart from a narrow stratum of
the aesthetically educated, who were familiar with
Werkbund simplicity and Bauhaus forms, to most
people, this device was a more or less exotic object.
[...] Its plain and linear design was simply unaccept-
able to the proverbial 'Joe Public'."[28]

Conceived according to the idea of good
form, the exclusive Braun devices were valued by
architects and designers. Almost all show apart-
ments at the Berlin Interbau trade fair (1957), which
were designed by international architects, featured
Braun devices that perfectly matched the modern
furniture.[29] Other institutions also honoured the
Braun designs. The New York Museum of Modern
Art, for example, showcased Braun radios in its 1958
exhibition of everyday objects. At the 1958 world exposition in Brussels, 16 devices were exhibited as
excellent examples of German design. Although the

Die exklusiven, der Idee der ‚Guten Form' verpflichteten Braun-Geräte wussten die Fachleute – Architekten und Interieurdesigner – zu schätzen. So wurden fast alle Musterwohnungen auf der Berliner Interbau-Ausstellung (1957) von verschiedenen internationalen Architekten mit Braun-Geräten ergänzt, die zu den modernen Möbeln sehr gut passten.[29] Darüber hinaus würdigten auch andere Institutionen die Braun-Entwürfe. Das Museum of Modern Art in New York etwa zeigte 1958 Braun-Radios anlässlich einer Ausstellung von Gebrauchsgegenständen. Auf der Weltausstellung in Brüssel 1958 wurden 16 Geräte als ausgezeichnete Beispiele deutscher Produktion ausgestellt. Obwohl die Radios nicht zum umsatzstärksten Segment der Firma Braun gehörten (das erfolgreichste Produkt ist bis heute der Elektrorasierer), trugen sie erheblich zum neuen Image der Firma bei und bilden heute ein wichtiges Kapitel der Designgeschichte. Die hier dargestellte Innovationskraft Brauns inspiriert noch heute zeitgenössische Produktdesigner und -designerinnen und äußert sich in der Gestaltung von neuen Produkten, wie zum Beispiel der Firma Apple (Abb. 4).

radios were not Braun's bestselling product lines (to this day, the most successful product is the Braun electric shaver), they considerably shaped the company's new image and represent a very important chapter in design history. Braun's innovative drive, as presented here, still inspires contemporary product designers and is expressed in the design of new products, for example in products from Apple (fig. 4).

1   Lenger, 2013, S. 507. Vgl. Beitrag von Breuer, Abb. 8, Reklame, „Konzerttruhe Komet", Kuba-Imperial, 1957.

2   Schmid, 1960, S. 67.

3   Gelsenkirchener Barock, 1991.

4   Nach seiner Schließung 1938 wurde der Werkbund im westlichen Teil Deutschlands 1947 reorganisiert und besteht bis heute. Er vereint Vertreter aus Industrie, Kunst und Handwerk.

5   Bill, 1957. Vgl. Hauffe, 2008, S. 117-118.

6   Der Deutsch-Amerikaner Hans Knoll gründete 1937 die „Hans G. Knoll Furniture Company", später „Knoll International", die bis heute existiert. Seiner Ehefrau Florence Knoll, Innenarchitektin und Möbeldesignerin, ist die Zusammenarbeit mit vielen namhaften Designern zu verdanken. Vgl. Günther, 1994, S. 127-133.

7   Wichmann, 1998, S. 47.

8   Neuausrichtung der Werbung, des Corporate Designs und des Firmenlogos. Die Schrift mit dem bereits verwendeten hochgezogenen „A" wurde 1952 von Wolfgang Schmittel mit Viertelkreisbögen abgerundet. Vgl. Wichmann, 1998, S. 47.

9   Polster, 2009, S. 33.

10 Zitiert nach Wichmann, 1998, S. 56-57.

11 Ebenda, S. 58.

12 Die HfG Ulm wurde auf Initiative von Inge Aicher-Scholl auf der Basis einer Volkshochschule gegründet. Getragen wurde sie zum größten Teil von der Geschwister-Scholl-Stiftung, die Inge Aicher-Scholl 1950 in Erinnerung an ihre Geschwister ins Leben rief. Die Schule bestand bis Dezember 1968.

13 Müller/Spitz, 2014, S. 69.

14 Ebenda.

1   Lenger, 2013: 507. See essay by Breuer, fig. 8, advertisement, "Konzerttruhe Komet", Kuba-Imperial 1957.

2   Schmid, 1960: 67.

3   Gelsenkirchener Barock, 1991.

4   After it was closed in 1938, the Werkbund was reestablished in West Germany in 1947. It still exists today, with members coming from the fields of industry, the fine arts and the crafts.

5   Bill, 1957. See also: Hauffe, 2008: 117-118.

6   German-American Hans Knoll founded the Hans G. Knoll Furniture Company in 1937, later renamed Knoll International. The company still exists today. The collaboration with many renowned designers is thanks to his wife Florence Knoll, an interior and furniture designer. See Günther, 1994: 127-133.

7   Wichmann, 1998: 47.

8   Re-design of advertising, corporate design and company logo. The typeface with the raised A was rounded off with quadrant arches in 1952 by Wolfgang Schmittel. See Wichmann, 1998: 47.

9   Polster, 2009: 33.

10 Quoted acc. to Wichmann, 1998: 56-57.

11 Ibid: 58.

12 The HfG Ulm was founded on the initiative of Inge Aicher-Scholl, based on the model of a 'Volkshochschule'. The school was mainly funded by the Geschwister Scholl Foundation, established in 1950 by Inge Aicher-Scholl to commemorate her siblings. The school existed until December 1968.

13 Müller/Spitz, 2014: 69.

14 Ibid.

15 Quoted acc. to Wichmann, 1998: 60.

16 Quoted acc. to Wichmann, 1998: 58.

15 Zit. nach Wichmann, 1998, S. 60.

16 Zit. nach ebenda, S. 58.

17 Ebenda, S. 60.

18 Hergestellt von Texas Instruments in Dallas, Texas und der Industrial Development Engineering Associates (I.D.E.A.) in Indianapolis, Indiana.

19 Polster, 2009, S. 74.

20 Braun-Produktkatalog „Radio-, Phono- und Fernsehgeräte", 1959, o. S.

21 Die Gestaltung der Schrift und grafischer Elemente übernahm in der Regel Otl Aicher. Vgl. von Seckendorff, 1985, S. 138.

22 Wichmann, 1998, S. 77.

23 Ebenda, S. 63.

24 Ebenda, S. 66.

25 Braun AG, 1960, o. S.

26 Braun kooperierte mit Knoll International und inszenierte seine Radios in Interieurs mit den Möbeln von Knoll.

27 Wichmann, 1998, S. 68.

28 Selle, 2007, S. 145.

29 Braun AG, 1960, o. S.

17 Ibid: 60.

18 Produced by Texas Instruments in Dallas, Texas and by Industrial Development Engineering Associates (I.D.E.A.) in Indianapolis, Indiana.

19 Polster, 2009: 74.

20 Braun product catalogue 'Radio-, Phono- und Fernsehgeräte', 1959, n. pag.

21 Otl Aicher usually designed the typography and graphic elements. See von Seckendorff, 1985:138.

22 Wichmann, 1998: 77.

23 Ibid: 63.

24 Ibid: 66.

25 Braun AG, 1960, n. pag.

26 Braun collaborated with Knoll International and showcased the Braun radios in interiors featuring Knoll furniture.

27 Wichmann, 1998: 68.

28 Selle, 2007: 145.

29 Braun AG, 1960, n. pag.

Radiodesign
im Space Age

Radio Design
in the Space Age

Isabel Brass

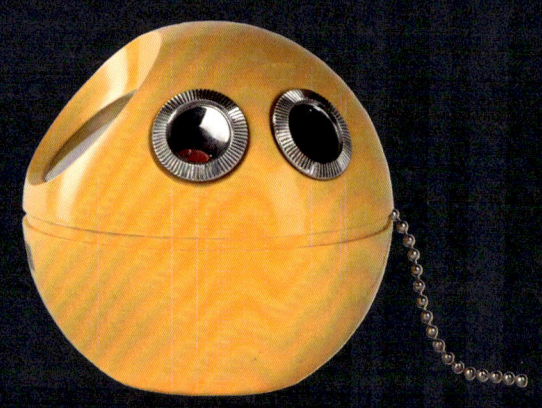

„Was heute noch wie ein Märchen klingt, kann morgen
Wirklichkeit sein. Hier ist ein Märchen von übermorgen…"

Theo Mezger alias W. G. Larsen

Diese erste Zeile aus dem Vorspann der deutschen Science
Fiction TV-Serie „Raumpatrouille – Die phantastischen Abenteuer des Raumschiffes Orion" kann als programmatisch für das sogenannte Space Age gelten. Sie verdeutlicht die Euphorie für die
in naher Zukunft erwartete Eroberung des Weltalls und den Fort-
schrittsglauben der 1960er und 1970er Jahre, als die Science
Fiction-Träume von der Realität regelrecht überholt wurden.

Das Raumfahrtzeitalter wurde 1957 mit der erfolgreichen
Entsendung des russischen Satelliten „Sputnik I" vom Daily
Express mit der Schlagzeile „Space Age is here" eingeläutet.[1] Es
entwickelte sich zwischen den Weltmächten in Ost und West ein
prestigeträchtiger Wettlauf ins All. Der Russe Juri Gagarin verließ

1 / Setfoto /Scene photo, Theo Mezger (Regie / Director),
„Raumpatrouille – Die phantastischen Abenteuer des Raumschiffes Orion",
Folge / Episode „Deserteure", (DE) 29.10.1966 © WDR/Bavaria.

"What sounds like a fairy tale today, may become reality tomorrow. Here's a fairy tale from the
day after tomorrow…"

Theo Mezger aka W.G. Larsen

The first line from the trailer for the German
science fiction TV series "Raumpatrouille – Die
phantastischen Abenteuer des Raumschiffs Orion"
(Space Patrol – The Fantastic Adventures of the
Spaceship Orion) can be considered programmat-
ic for what is generally referred to as the 'space age'.
The statement highlights the euphoria for the then-
soon-to-be-expected conquest of space and the
faith in progress that dominated the 1960s and '70s,
a time when science-fiction dreams were quite lit-
erally surpassed by reality.

Referring to the successful 1957 launch of the
Russian "Sputnik I" satellite, The Daily Express her-
alded the new era with the headline "Space Age is
here".[1] What followed was a high-profile race for su-
premacy in spaceflight between the eastern and
western superpowers. In 1961, the Russian Yuri
Gagarin was the first human to leave the Earth's at-
mosphere, relaying his breathtaking experience to
an enthralled world audience.[2] One year later, pres-
ident John F. Kennedy promised to send astronauts
to the moon by the end of decade. The Apollo
11 mission's 1969 moon landing represented the climax of the space age, with millions of people
around the world sitting in front of their TV or radio
sets to witness the event. These rapid developments
created a notion of boundless possibility and
sparked ideas about colonising the Moon and Mars.[3]

Filmmakers picked up on the unfolding fan-
tasies and translated them into TV series and feature
films. The first episode of the above mentioned Ger-

1961 als erster Mensch die Erdatmosphäre und berichtete der gespannten Weltöffentlichkeit von seinen atemberaubenden Eindrücken.[2] Ein Jahr später gab Präsident John F. Kennedy das Versprechen, bis Ende des Jahrzehnts Astronauten auf den Mond zu entsenden. Die Mondlandung der „Apollo 11" Mission 1969 stellte den Höhepunkt des Space Age dar. Ein Ereignis, das von Millionen Menschen weltweit an Radio- und Fernsehgeräten verfolgt wurde. Diese rasanten Entwicklungen erzeugten ein Gefühl der unbegrenzten Möglichkeiten und es wurde bereits über Mond- und Marskolonien spekuliert.[3]

Filmemacher griffen die sich entspinnenden Fantasien auf und setzten sie in Serien und Spielfilmen um. Für Deutschland wurde bereits die TV-Serie „Raumpatrouille Orion" genannt, die im September 1966 erstmalig ausgestrahlt wurde, im selben Monat wie die bis heute erfolgreiche Serie „Star Trek" in den USA. Beide zeigen das optimistische Bild einer friedlich vereinten Weltbevölkerung und der Erforschung des Weltraums mithilfe ausgefeilter Technologien. Der Regisseur Stanley Kubrick hingegen zeichnete im erfolgreichsten Kinofilm des Jahres 1968 „2001: A Space Odyssey" ein düsteres und enigmatisches Bild des Raumfahrtzeitalters. Er behandelt unter anderem Fragestellungen im Zusammenhang mit Robotern und künstlicher Intelligenz. Diese Serien und Filme mit ihren Requisiten und Sets liefern eine Grundlage für die Begeisterung der zeitgenössischen Architekten, Modeschöpfer und Designer für den futuristischen Space Age-Look (Abb. 1).[4]

man TV series "Raumpatrouille Orion" was screened in September 1966, in the same month as the US "Star Trek" series that still enjoys great popularity today. Both series featured optimistic ideas of a peacefully united Earth population and of using sophisticated technologies for space exploration. Director Stanley Kubrick, on the other hand, portrayed the space age in dark and enigmatic pictures in his 1968 film "2001: A Space Odyssey", the most successful motion picture of the year. Among other subjects, the film focused on problems related to robots and artificial intelligence. With their props and sets, these series and films delivered the blueprint for the futuristic space-age look favoured by architects, fashion designers and product designers of the time (fig. 1).[4]

Real-world space travel and the hypothetical colonisation of alien planets involved highly challenging requirements in terms of independent living in the smallest of spaces. One expected that, by the year 2000, residential architecture and home design would have undergone radical changes.[5] Accordingly, architects and designers developed new and versatile solutions, presented as conceptual living systems, which included the 1969 "Visiona I" by Joe Colombo for Bayer AG. As part of the 1969 Cologne furniture fair, Colombo's interior design of an exhibition ship not only presented a utopian model of living (Wohnutopie)[6] but also showcased numerous applications of plastic materials which were highly popular in the 1960s (fig. 2).[7] The use of smooth plastic surfaces demonstrated the technological progress in the areas of materials and finish, thus also indicating a positive attitude towards machine-based production. The "Visiona I" interior design concept was defined by smooth surfaces and volumes. None of the windows was used to create a

2 / Modell / Model „Visiona I", Joe Colombo, Bayer AG, Köln / Cologne (DE) 1969
© Bayer AG, Corporate History & Archives.

Die reale Raumfahrt und die hypothetische Kolonisierung fremder Planeten stellten hohe Anforderungen an ein möglichst autonomes Wohnen auf engstem Raum. Man erwartete eine radikale Veränderung des Wohnens bis zum Jahr 2000.[5] Architekten und Designer entwickelten in diesem Sinne neue, flexible Lösungen, die sie in Wohnkonzepten präsentierten. Dazu zählt „Visiona I" von Joe Colombo im Auftrag der Bayer AG von 1969. Colombos Ausstattung eines Ausstellungsschiffes zeigte im Kontext der damaligen Kölner Möbelmesse nicht nur eine „Wohnutopie"[6], sondern auch die vielfältigen Einsatzmöglichkeiten von Kunststoffen, die in den 1960er Jahren eine Blütezeit erlebten (Abb. 2).[7] Die Verwendung von glatten Kunststoffoberflächen demonstrierte den technischen Fortschritt in Material und Verarbeitung und verwies dadurch auf die positiv konnotierte maschinelle Produktion. Die Raumausstattung der „Visiona I" ist geprägt von solch glatten Oberflächen und stereometrischen Formen. Durch kein Fenster wird eine Verbindung zur Außenwelt hergestellt und so verwandelt sich der Raum in eine künstlich beleuchtete Weltraumkapsel, die ihrem Bewohner viele Annehmlichkeiten bietet. Hinter kugeligen, raumgreifenden Objekten verbergen sich eine Schlaf- und Naßzelle sowie eine abgeschlos-

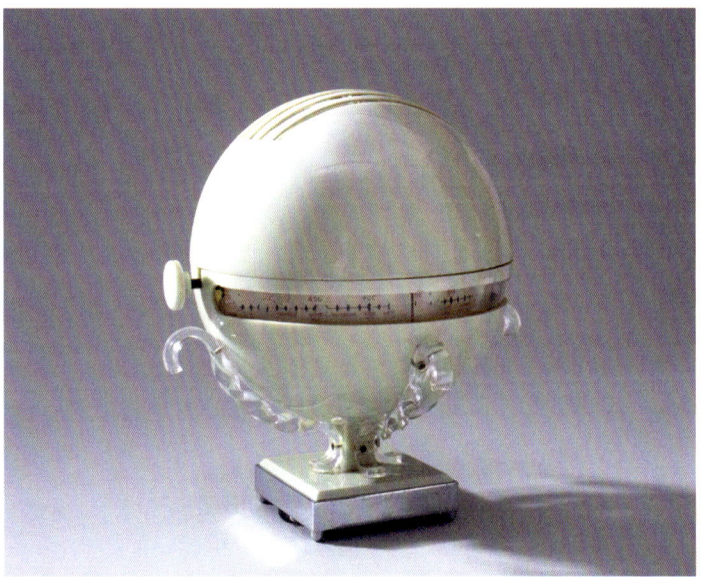

3 / Venus, Champion Electric, Newheaven und / and Seaford (GB), 1947, Science Museum London © Science & Society.

connection to the outside world, thus transforming the interior space into an artificially lit space capsule that offered many amenities to its users. Large spherical enclosures served as sleeping pod, wet room and self-contained kitchen unit, while a central lounging area featuring a ceiling-mounted TV invited users to rest and relax. Sean Topham interprets this division of space and the emphasis given to individual functions as references to the dream of a machine-run world that gives people the freedom to relax and indulge in media consumption.[8] The media – in this case radio and television – re-established the connection to the outside world, albeit indirectly.

Inspired by the historical events of the time and their documentation in countless photos and TV images, designers developed a range of colours and forms based on white astronauts' helmets and silver-coloured space capsules set against the backdrop of the universe and its many celestial bodies. These influences can be found not only in film sets and utopian models of living, but also very tangibly in furniture and product design. In the consumer electronics industry, radio and TV sets were designed using space-age attributes.[9] An early predecessor is the 1947 "Venus" radio by the British Champion Electric Corporation, which already anticipated the characteristics of the space age radios from the 1960s and early 1970s (fig. 3). The spherical "Venus" radio is supported by four twisted clear acrylic arms, which are fixed to a square base. The two halves of the white plastic sphere are separated by an acrylic strip through which the tuning dial is visible. The slits for the speakers are located at the top of the upper hemisphere. Two knobs located at opposite sides are presumably used for volume control and tuning. The plastic is translucent so that the

sene Kücheneinheit, während eine zentrale Liegemöglichkeit mit einem eingebauten Fernseher unter der Decke zum Verweilen einlädt. Sean Topham sieht in dieser Aufteilung und Gewichtung der verschiedenen Funktionen einen Hinweis für den Traum von einer maschinendominierten Welt, die dem Menschen die Freiheit gibt sich zu entspannen und dem Medienkonsum hinzugeben.[8] Durch diese Medien, namentlich Radio und Fernsehen, wird der Bezug zur Außenwelt mittelbar wiederhergestellt.

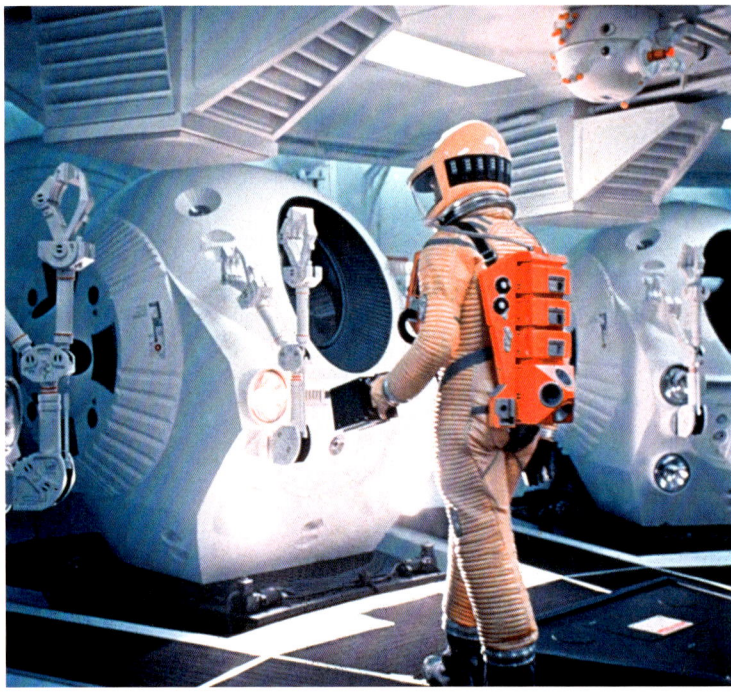

4 / Setfoto / Scene photo, Stanley Kubrick (Regie / Director), MGM, „2001: A Space Odyssey" (UK; US) 1968 © Kobal.

Unter dem Eindruck der historischen Ereignisse und ihrer Dokumentation in zahlreichen Fotografien und Fernsehbildern entwickelten Designer ein Farb- und Formenspektrum, das sich an den weißen Astronautenhelmen und silbernen Raumfahrtkapseln vor dem Hintergrund des Weltalls und der Himmelskörper orientierte. Diese Einflüsse lassen sich nicht nur an Filmsets und utopischen Wohnkonzepten, sondern auch konkret im

radio is illuminated from within when in use. Form and colour are not reminiscent of astronauts' helmets or space capsules; rather the radio invokes images of shining celestial bodies, thus expressing the postwar era's fascination with outer space.[10]

Also reminiscent of planets are the "R 70, Panapet" spherical radios by Matsushita Electric Industrial[11], designed in 1970 (cat.no. 129). These small, portable radios were produced with a plastic casing available in six bold colours and feature flat, chrome-effect dial controls for volume and tuning. The black circular tuning dial with its small spherical red pointer is embedded in the circular frame of the truncated casing.[12] The radio has a flat-based bottom, but it can also be suspended with a metal bead chain and ring to create an impression of zero gravity. Here too, the spherical form and the option to suspend the radio are reminiscent of a planet, satellite or capsule moving through space. The design also invokes images of the space capsule from "2001: A Space Odyssey" and of the threatening red camera 'eye' of HAL, the space ship´s onboard computer (fig. 4).

Using a larger spherical form than the above-described products, Thilo Oerke's design "Vision 2000" audio system was produced by Rosita Tonmöbel in the early 1970s (fig. 5): a white semi-spherical form is mounted on a chrome-plated trumpet-shaped base, with a tape deck, a radio and a control unit set in a black surface. A dome-like bowl made of transparent plastic serves as cover and, when closed, completes the volume. With this type of cover, the designer created a space-saving solution that visually alludes to astronauts' helmets with visors.

Möbel- und Produktdesign nachweisen. Im Bereich der Unterhaltungselektronik wurden Radio- und Fernsehgeräte mit Merkmalen des Space Age entworfen.[9] Ein früher Vorläufer ist das Modell „Venus" von 1947 der britischen Firma Champion Electric, das bereits alle Charakteristika der Space Age Radios der 1960er und frühen 1970er vorweg nimmt (Abb. 3). Auf einer Sockelplatte sind vier mehrfach in sich gedrehte Halter aus farblosem Acryl befestigt auf denen das kugelförmige Radio ruht. Die Hemisphären der weißen Kunststoffkugel werden durch ein Acrylband getrennt, hinter dem die Skala sichtbar wird. Im oberen Teil der Kugel befinden sich Lautsprecherschlitze. Über zwei Drehregler an gegenüberliegenden Seiten werden vermutlich Lautstärke und Frequenz eingestellt. Der verwendete Kunststoff ist durchscheinend, so dass das Radio im Betrieb von innen heraus leuchtet. Dies erinnert in Form und Farbe nicht an Astronautenhelme oder Raumfahrtkapseln, sondern an leuchtende Himmelskörper und verdeutlicht somit die Weltraumbegeisterung der Nachkriegszeit.[10]

Die Kugelradios „R 70, Panapet" von Matsushita Electric Industrial[11], entworfen 1970, erinnern ebenfalls an Planeten (Kat. Nr. 129). Die kleinen tragbaren Geräte wurden in sechs intensiven Farbvarianten aus Kunststoff produziert und sind mit silberfarbenen flachen Drehreglern für Lautstärke und Frequenz versehen. Durch einen kreisrunden Anschnitt wird die schwarze ringförmig angeordnete Skala mit einer kleinen roten Kugel als Zeiger sichtbar.[12] Neben einer abgeflachten Standfläche verfügt das Gerät über eine metallene Kugelkette mit Ring, an der das Gerät aufgehängt und so der Eindruck von Schwerelosigkeit erzeugt werden kann. Auch hier erinnern die Kugelform und die Möglichkeit zum Herabhängen des Radios an einen Planeten, Satelliten oder eine Kapsel, die sich durch den Raum bewegt. Assoziationen mit der Raumfahrtkapsel aus „2001: A Space Odyssey" und dem bedrohlichen roten ‚Kamera-Auge' des Boardcomputers „HAL" werden ebenfalls geweckt (Abb. 4).

Der Entwurf von Thilo Oerke für das Kombigerät „Vision 2000" wurde Anfang der 1970er von der Firma Rosita Tonmöbel aufgenommen und setzt die Kugelform in größeren Dimensionen um (Abb. 5). Auf einem verchromten Trompetenfuß ist eine weiße Halbkugel aus Kunststoff aufgesetzt. In die schwarze ebene Fläche sind ein Kassettendeck, ein Radio und eine Steuereinheit

Die Raumfahrt bringt neue Impulse

Supermodernes Styling und weichgerundete Formen werden durch Kunststoffe realisiert.

„Commander Luxus" und „Vision 2000" sind modische Varianten der Kompaktanlagen. Form und Technik sind auf dem neuesten Stand.

Die neuen Materialien erfordern zusätzliche Fertigungsverfahren. ROSITA geht am Markt nicht vorbei.

5 / F. J. Stütner (Konzept / Concept):
„Musik wohnlich verpackt. 25 Jahre Rosita Tonmöbel",
Paderborn-Schloß Neuhaus (DE) 1979.

The motif of the astronaut's helmet is even more obvious in the "2004 Prinz Sound-Stereo" radio cassette recorder (cat.no. 136). A large truncated sphere rests on a short base ring that could be interpreted as a neck. The black truncated surface, in which the control elements are embedded, is suggestive of a helmet's visor. With this design, the device seems to be slightly inclined upwards as if the helmet were looking up towards the user. At the sides, where the wearer's ears would be, are the speakers, from which the stereo sound alluded to in the product's name emerges. The speakers

eingelassen. Eine kuppelartige Schale aus durchsichtigem Kunststoff dient als Abdeckung und vervollständigt in geschlossenem Zustand die stereometrische Grundform. Durch diese Art der Abdeckung wurde eine sehr platzsparende Lösung gefunden und zugleich optisch die Nähe zu den Astronautenhelmen mit Visier erzeugt.

6 / Reklame / Advertisment, „The new shape of sound", Weltron, Durham, North Carolina (US) ca. 1973
© Thierry Didion.

feature disc-shaped covers with concentrically arranged perforations. The retractable handle is reminiscent of a comb or brush; a metal loop next to the handle can be used to suspend the device. The MAKK collection includes the white design of the "2004 Prinz Sound-Stereo", which was also available in yellow and orange. Matching spherical speakers were offered in the same colours. Based on the first "2001" model, Weltron[13] produced the entire 2000 range using the same visual design, with individual models only differentiated in terms of technical specification. Using the slogan 'The new shape of sound', in 1973, Weltron launched its new models at the Chicago Consumer Electronics Show (fig. 6).[14] Like the "Venus" radio from the 1940s, the products' names alluded to the space age by including numbers from the 2000-range. The first "2001" model, for example, harked back to Stanley Kubrick's film[15] while simultaneously pointing towards the forthcoming turn of the millennium, which is also referenced in Joe Colombo's utopian models of living.

The same is true for the "Vision 2000" audio furniture by Rosita Tonmöbel. Besides the prominent spherical form, which is also known from furniture design,[16] there are also rectangular forms: Rosita Tonmöbel used edgy shapes for its compact audio system, whose look and name are reminiscent of the control panel in a science fiction spaceship (fig. 5). The "Commander Luxus" design features an L-shape supported by a chrome-plated base. A turntable and a tape deck with a clear plastic cover are positioned at the horizontal plane.[17] Controls, radio and speakers are embedded in the vertical plane. The tuning dial is particularly striking with its backlit vertical-rectangular black surface illuminating the display's green lettering. In a publication celebrating the company's 25th anniversary,

Das Motiv des Astronautenhelms wird noch deutlicher im Radiorekorder Modell „2004, Prinz Sound-Stereo" umgesetzt (Kat. Nr. 136). Auf einem kurzen Standring, der als Hals gesehen werden kann, ruht eine voluminöse Kugel, die im oberen Bereich angeschnitten ist. Durch diese schwarze Schnittfläche, in welche die Bedienelemente eingelassen sind, wird der Sichtbereich des Helms markiert. Das Gerät wirkt dadurch leicht nach oben geneigt, als würde der Helm den Radiohörer von unten her anschauen. Seitlich der Fläche, wo sich die Ohren des Helmträgers befänden, sind die Lautsprecher installiert, aus denen der im Gerätenamen angegebene Stereo-Sound erklingt. Sie werden von je einem Kreis mit konzentrisch angeordneten Löchern verkleidet. Wie ein Kamm mutet der versenkbare Tragegriff an, neben dem eine Öse zur Aufhängung angebracht ist. In der Sammlung des MAKK befindet sich die weiße Farbvariante des Modells „2004", darüber hinaus waren eine gelbe und eine orange Variante erhältlich. Passend zum Gerät wurden kugelförmige Lautsprecher in denselben Farben angeboten. Ausgehend vom ersten Modell „2001" produzierte die Firma Weltron[13] eine ganze „2000"er-Serie in diesem Design, die sich lediglich in ihren technischen Funktionen unterscheidet. Unter dem Slogan „The new shape of sound" stellte Weltron 1973 seine neuen Modelle auf der Consumer Electronics Show in Chicago vor (Abb. 6).[14] Wie schon die „Venus" aus den 1940er Jahren, so verweist auch in diesem Fall der Modellname durch die 2000er-Zahlen auf das Space Age. Zum einen wegen des ersten Modells 2001, das Bezug auf Stanley Kubricks Filmtitel nimmt,[15] zum anderen deutet es auf die bevorstehende Jahrtausendwende hin, in die auch die Wohnutopien eines Joe Colombo verweisen.

Dies gilt ebenso für „Vision 2000" von Rosita Tonmöbel. Denn neben der prominenten Kugelform, die auch aus dem Möbeldesign bekannt ist,[16] werden alternativ rechteckige Formen eingesetzt. Rosita Tonmöbel kombiniert kantige Formen zu einer Kompaktanlage, die optisch und namentlich an die Steuerkonsole eines Science Fiction-Raumschiffs erinnert (Abb. 5). Der „Commander Luxus" steht auf einem verchromten Standfuß und bildet eine L-Form. Auf der horizontalen Ebene befinden sich ein Plattenspieler und ein Kassettendeck unter einer durchscheinenden Plastikhaube.[17] Auf der vertikalen Ebene sind die Bedienelemente, das Radio und die Lautsprecheröffnungen eingelassen. Besonders auffällig an dieser Stelle ist die Skala auf einer hoch-

Rosita Tonmöbel refers to spaceflight as the inspiration behind the "supermodern styling" of the "Commander Luxus" and "Vision 2000" designs.[18]

In 1965, Achille and Pier Giacomo Castiglioni designed the "RR 126 – OF-ST" stereo system for Italian consumer electronics company Brionvega (cat.no. 120). Unlike the above-mentioned products, the Castiglioni design uses painted wood instead of plastic. The version with a clear varnish finish even retains the natural wood look. A rectangular case featuring chrome-effect controls and two black tuning dials rests on an aluminium base with castors. The dials are shaped like segmental arches, each one of which is positioned above a control rotary. Below those is a centred panel of push buttons and another row of rotary controls. The arrangement is reminiscent of a face, with the base looking like neck and shoulders, and evokes associations of an expressive 'music robot'.[19] The 'music robot's' speakers are constructed as modular elements, which can either be hooked to the case's sides for stereo sound or put next to each other on top of the case for mono sound. When the mono set-up is used, the turntable disappears into the casing of one of the speakers. With its 'facial expression', the system has a creature-like appearance, thus referencing discourses on artificial intelligence and on machines as servants for humans. Both of these topics were addressed in the space age. Due to its useful features and outstanding design, the "RR 126" soon became a classic product of its time.[20]

Together with renowned designers like Richard Sapper, Marco Zanuso and Mario Bellini, during the 1960s and 70s, Brionvega continued to create further design highlights in the consumer electronics sector. In the 1960s, the company

rechteckigen schwarzen Platte, die aus dem Inneren angestrahlt die grünen Lettern leuchten lässt. In ihrer Festschrift zum 25-jährigen Firmenjubiläum nennt Rosita Tonmöbel die Raumfahrt als Impulsgeber für das „supermoderne Styling" des „Commander Luxus" und des „Vision 2000".[18]

Achille und Pier Giacomo Castiglioni entwarfen 1965 für das italienische Elektronikunternehmen Brionvega die Stereoanlage Modell „RR 126 – OF-ST" (Kat.Nr. 120). Im Unterschied zu den bisher vorgestellten Geräten ist dieses nicht aus Kunststoff, sondern aus lackiertem Holz gefertigt. In der Variante mit Klarlack behält es sogar seine Holzsichtigkeit. Auf einem Aluminiumfuß mit Rollen ruht ein längsrechteckiger Quader mit silberfarbenen Bedienelementen und zwei schwarzen Skalen. Die Skalen sind als Segmentbögen über je einem Drehregler angebracht, zwischen denen eine Leiste aus Drucktasten und eine Reihe weiterer Drehregler eingefügt ist. Diese Anordnung erweckt den Eindruck eines Gesichts und lässt den Standfuß wie Hals und Schultern erscheinen. Assoziationen mit einem ausdrucksstarken ‚Musikroboter' kommen auf.[19] Die Lautsprecher des ‚Musikroboters' sind modular angelegt, denn für den Stereosound können sie an die Schmalseiten des Gerätes gehängt, für einen Monosound nebeneinander auf das Gerät gestellt werden, wobei der Plattenspieler im Gehäuse eines der Lautsprecher verschwindet. Die Stereoanlage erhält durch ihre ‚Gesichtszüge' etwas Wesenhaftes und nimmt dadurch Bezug auf Diskurse zur künstlichen Intelligenz und der Maschine als Diener des Menschen. Beide Themen wurden im Rahmen des Space Age behandelt. Wegen seiner praktischen Qualität und seinem außergewöhnlichen Design wurde das Modell „RR 126" bald zu einem Klassiker seiner Zeit.[20]

Brionvega kreierte zusammen mit namhaften Designern wie Richard Sapper, Marco Zanuso und Mario Bellini in den 1960er und 1970er Jahren weitere Design-Highlights der Unterhaltungselektronik. In den 1960er Jahren reiste der Sohn des Firmengründers Ennio Brion zusammen mit Zanuso und Sapper in die USA und besichtigte die Zentren der Weltraumforschung.[21] Eine Beeinflussung ihrer Entwürfe ist also durchaus wahrscheinlich. Auch weisen die Brionvega-Geräte dieser Zeit die oben dargelegten Merkmale des Space Age-Designs auf. Das tragbare Klappradio „TS 502" (Kat.Nr. 117) und die Stereoanlage „Totem" (Kat.Nr. 132) werden aus stereometrischen Körpern gebildet, sie

founder's son, Ennio Brion, went to the US together with Zanuso and Sapper to visit the space research centres[21]. This visit will likely have influenced their designs. The Brionvega products from the time also feature the above-described space age characteristics. The portable and foldable "TS 502" radio (cat.no. 117) and the "Totem" stereo system (cat.no. 132) both have cuboid forms and are made from plastic and painted wood with smooth surfaces. The predominant colours are white, black and silver as well as the base colours red, yellow and blue. The same is true for Verner Panton's system for Wega (cat.no. 128) and for the "R 72, Toot-a-loop" bracelet radio (cat. no. 125) by Matsushita Electric Industrial. However, these last two products cannot be clearly assigned to the space age as the characteristics of this particular design style are not used in such a way as to create an allusion to space travel. This is not unusual as the typical space age forms, colours and materials also appeared in other design styles of the time and the boundaries between different styles were often fluid.[22]

1 Pincher, 1957: 1.

2 See Crosscurrents Press, 1961.

3 See Topham, 2003: 42.

4 See Garner, 2003: 98 -100.

5 Topham, 2003: 104.

6 Schepers, 1998: 26.

7 Topham, 2003: 67.

8 Ibid: 112.

9 Space age design TV sets included in the MAKK collection: Nivico from JVC Jap, Yokohama, Japan, 1965-75 (inv.no. K 691); Nivico 3240 GM TV Videosphere from JVC Jap, Yokohama, Japan, 1973 (inv.no. K 1218 W).

10 Attwood, 1997: 23.

sind aus Kunststoff und lackiertem Holz mit glatter Oberfläche hergestellt und vorwiegend in weiß, schwarz, silberfarben und den Grundfarben rot, gelb, blau gehalten. Dies gilt ebenso für das Kombigerät von Verner Panton für Wega (Kat.Nr. 128) und das Armreif-Radio „R 72, Toot-a-loop" (Kat.Nr. 125) von Matsushita Electric Industrial. Diese Modelle sind nicht eindeutig dem Space Age zuzuordnen, da sich in ihnen die Merkmale dieser Design-richtung nicht zu einer Allusion der Raumfahrt verdichten. Dies ist nicht ungewöhnlich, denn die typischen Space Age-Formen, Farben und Materialien traten ebenfalls in anderen zeitgleichen Design-Strömungen auf und die Grenzen waren oft fließend.[22]

1   Pincher, 1957, S. 1.

2   Vgl. Crosscurrents Press, 1961.

3   Vgl. Topham, 2003, S. 42.

4   Vgl. Garner, 2003, S. 98-100.

5   Topham, 2003, S. 104.

6   Schepers, 1998, S. 26.

7   Topham, 2003, S. 67.

8   Ebd. S. 112.

9   Fernsehgeräte im Space Age-Design in den Sammlungen des MAKK: JVC Jap, Modell "Nivico", Yokohama, Japan, 1965-75 (Inv.Nr. K 691); JVC Jap, Modell "Nivico 3240 GM TV Videosphere", Yokohama, Japan, 1973 (Inv.Nr. K 1218 W).

10 Attwood, 1997, S. 23.

11 Matsushita Electric Industrial (MEI) wurde 1918 von Kōnosuke Matsushita gegründet und firmiert heute als „Panasonic Corporation". MEI vertrieb seine Elektronikgeräte in Deutschland unter anderem mit den Markennamen „Panasonic" und „National".

12 Bedient man den Frequenzregler, so dreht sich die Scheibe mit der Skala, nicht aber der Zeiger.

13 Der Hersteller Weltron Co. Inc. war ein amerikanisch-japanisches Joint Venture mit Hauptsitz in Durham, North Carolina.

14 Joe, 1972, S. 32.

15 Breuer, 2001, S. 186.

16 Vgl. Eero Aarnios „Ball Chair" von 1963 auf einem Standfuß und seinen herabhängenden „Bubble Chair" von 1968.

17 Die technischen Geräte wurden von der Elektronikfirma Philips geliefert.

18 Stütner, 1979, o.S.

19 Carugati, 2003, S. 75 und 79.

20 Ebd., S. 75.

21 Ebd., S. 54.

22 Breuer, 2001, S. 189.

11 Matsushita Electric Industrial (MEI) was established in 1918 by Kōnosuke Matsushita. Today, the company is known as Panasonic Corporation. MEI sold its electronic products in Germany using, among others, the brand names Panasonic and National.

12 When using the tuning control, the disc-shaped dial display rotates while the pointer does not move.

13 Weltron Co. Inc. was a US-Japanese joint venture headquartered in Durham, North Carolina.

14 Joe, 1972: 32.

15 Breuer, 2001: 186.

16 See Eero Aarnio's 1963 "Ball Chair" mounted on a swivel base and his 1968 suspended "Bubble Chair".

17 The technical appliances were supplied by electronics manufacturer Philips.

18 Stütner, 1979, n. pag.

19 Carugati, 2003: 75 and 79.

20 Ibid.: 75.

21 Ibid.: 54.

22 Breuer, 2001: 189.

Von ganz groß bis ganz klein –
Die 1980er bis heute

From XXL to XXS: The 1980s to the Present

Theresa Nisters

„Less is a bore"[1] lautet das Motto, unter dem Robert Venturi 1969 dem puristischen Stilideal des 20. Jahrhunderts den Kampf ansagt. War der Modernismus bis dahin stilprägend und hatte die Prinzipien der ‚guten Form', mit Klarheit, Abstraktion und Einfachheit definiert, gilt es ab den 1970er Jahren, die Zeichen umzukehren.

1 / Carlton, Ettore Sottsass, Memphis s.r.l., Pregnana Milanese (IT) 1981, Inv.Nr. Ov 190, MAKK © Foto / Photo: Rheinisches Bildarchiv Köln / Cologne, Marion Mennicken.

In den internationalen Designhochburgen Mailand, Paris, Berlin und London bricht eine neue, selbstbewusste Generation an Gestaltern mit dem kühlen Rationalismus. Innovatives Design setzt sich zu den Vorgängern ab, indem es haptische Sinnlichkeit, Ornament, Farbe und freie Form in einem möglichst singulären Entwurf realisiert. Dabei werden nicht nur die Grenzen zu Kitsch und Alltagsbanalität ohne Hemmung überschritten, son-

"Less is a bore"[1]: with his 1969 statement, Robert Venturi challenged the purist style ideal of the 20th century. While, until then, modernism had been the major stylistic influence, defining the principles of 'good form' with clarity, abstraction and simplicity, the 1970s saw a reversal in attitude.

In the international design strongholds of Milan, Paris, Berlin and London, a new and confident generation of designers broke with the ideals of cool rationalism. Innovative design set itself apart from preceding movements by implementing tactile sensuality, ornament, colour and free forms in individual designs. These new designers were not afraid to even venture into kitsch and mundane banality and neither did they hesitate to confront historical styles and topple conventional ideas about the value of materials.[2] In 1981, the show by the Memphis design group at the Milan furniture fair rocked the international design scene to its core. The group's loud and humorous designs looked like they were cobbled together from plywood and covered with a patterned 'skin' (fig. 1).

The flashy playfulness displayed in some of the new designs did, however, not result from thoughtless creative urges. Rather, what was mirrored in this quality was historical awareness and engagement with theory. The modernist paradigm, according to which an object's form had to follow its function, was being subjected to revision. In 1981, Daniel Weil put the electronic components of a radio receiver in a clear plastic cover adorned with colourful graphic patterns and the word 'radio'. The aptly named "Radio Bag" (fig. 2) undermined the conventional idea of a radio as a closed box. By having its technical components exposed in a PVC skin, the object challenges the actual potential of

dern auch historische Stile konfrontiert und konventionelle Materialwerte umgeworfen.[2] 1981 löst der Auftritt der Designgruppe „Memphis" auf der Mailänder Möbelmesse einen regelrechten Schock in internationalen Designkreisen aus. Ihre kreischend bunten, humorvollen Entwürfe scheinen wie aus Sperrholzteilen wild zusammengehämmert und sind mit einer ‚Haut' aus gemustertem Laminat überzogen (Abb. 1).

Die zum Teil schrille Verspieltheit der neuen Entwürfe resultiert jedoch nicht aus unreflektiertem Gestaltungswillen, sondern spiegelt historisches Bewusstsein und theoretische Auseinandersetzung wider. Das modernistische Paradigma, nach welchem die Form eines Gegenstandes dessen Funktion zu folgen hat, wird einer Revision unterzogen. Daniel Weil steckt 1981 die elektronischen Bauteile eines Rundfunkempfängers in eine transparente Plastikhülle, die mit farbigen, grafischen Mustern und dem Schriftzug „Radio" geziert ist. „Radio Bag" (Abb. 2), so der unverhohlene Titel, unterläuft die herkömmliche Vorstellung eines Radioapparats als geschlossenem Kasten. Indem die technischen Bestandteile in der PVC-Hülle sichtbar bleiben, hinterfragt das Objekt das tatsächliche Potenzial des Prinzips „form follows function". Der abstrakte Dekor sowie die Verwendung von Buchstaben auf der Plastiktasche erinnern zudem an Bildcollagen des Kubismus, einer künstlerischen Strömung zu Beginn des 20. Jahrhunderts, die das Verhältnis von illusionistischem Bildraum, materiellem Bildträger und Rahmung hinterfragte und damit bisherige Kategorien der Malerei auf den Kopf stellte.

In der zweiten Hälfte des 20. Jahrhunderts nimmt eine neue künstlerische Strömung erheblichen Einfluss auf die allgemeine Geisteshaltung und hat spezifische Auswirkungen auf das Produktdesign: Von England ausgehend macht sich die Pop-Kultur auf internationalen Eroberungszug.[3] Sie prägt vor allem eine junge Käuferschicht, die im Zeitalter der Massenproduktion Konsum zelebriert und entsprechend dem Slogan „I like things"[4] ihres berühmten Wortführers Andy Warhol Objektkult betreibt. Firmen nutzen den Warenfetisch in ihren Marktstrategien. Radiogeräte, die als Werbegeschenke fabriziert werden, geben in Form von Miniaturskulpturen und glänzender trompe-l'œil-Optik keinerlei Hinweis auf ihre Funktion. So beherbergt auch die Benzinzapfsäule des Feuerzeugproduzenten Marksman

the 'form follows function' principle. Furthermore, both the abstract décor and the use of letters on the plastic bag are reminiscent of cubist collages, an art movement at the beginning of the 20th century that questioned the relationship between the illusionistic pictorial space, the material that carries the image and the image's frame, thus turning the then-conventional categories of painting upside down.

During the second half of the 20th century a new art movement significantly shaped the world of culture and had a specific effect on product design: originating in England, pop culture started to take the world by storm.[3] It was most influential among young consumers who, in the age of mass production, celebrated consumption and indulged in an object cult, inspired by the statement "I like things"[4], which was made by one of the generation's most prominent spokespeople, Andy Warhol. Companies used the commodity fetish in their marketing strategies. Produced as promotional giveaways, radios came in the form of miniature sculptures featuring trompe-l'œil looks that did not give away any clues as to their function. The petrol pump by lighter manufacturer Marksman Polyflame, for instance, also housed a clock radio. This witty PR gag was given additional value by the tag 'Collector's Edition', which designated the lighter as a valuable and rare collector's item (cat.no. 147).

The object was given symbolic status: as an external symbol it became a visible expression of unique individuality while at the same time identifying its owner as a member of a particular group.

The new self-confidence of consumers was met with a new kind of product development. For the Japanese consumer electronics manufacturer

2 / Radio Bag, Daniel Weil, APEX, Tokio (JP) 1981-83
© Pentagram Design Ltd.

Polyflame einen Radiowecker. Aufgewertet wird dieser witzige PR-Kniff durch den Titelzusatz „Collector's Edition", der ihn als wertvolle Rarität und Sammelobjekt auszeichnet (Kat.Nr. 147).

Das Objekt erhält symbolischen Status: als äußeres Zeichen ist es der für alle sichtbare Ausdruck unverwechselbarer Individualität – und wird zugleich Identifikationsfaktor von Gruppenzugehörigkeit.

Dem neuen Selbstbewusstsein des Käufers wird durch eine neue Form der Produktentwicklung entsprochen. Für den japanischen Elektronikhersteller Sony überwiegt das technische Innovationsinteresse vor ausgeklügeltem Design.[5] Zu Beginn der 1980er Jahre schafft er es jedoch zu großem Markterfolg.

Sony, technical innovation was more important than sophisticated design.[5] Nevertheless, by the beginning of the 1980s, Sony had achieved great market success. Products were now developed by matching technologies to consumer profiles, like in Sony's "ICF-7600 A", launched in 1982 (cat.no. 142).[6] At the end of the 1950s, with the emergence of the transistor radio, people were no longer confined to their living rooms when listening to the radio. During the 1970s, the technology was refined and shrunk to fit in ever-smaller spaces. The resulting compact systems, equipped with practical handles, could easily be taken on holidays. The most striking features of the Japanese "ICF-7600 A" design are its pocketbook size and minimal weight.

However, it was the New York Hip Hop scene that helped the portable radio with integrated cassette deck or CD player to reach 'true greatness' and fame during the 1980s. The 'ghetto blaster' or 'boom box' conquered the streets and became both a symbol of a subculture's freedom and a youth cult object.[7] Providing the soundtrack to the breakdance movement, the ghetto blaster became a recurrent motif in graffiti art and on record covers. In 1985, Virgin Interactive even published a computer game named after this type of radio-cassette player. While being one of the smaller examples of its kind, Sony's 1987 "CFS-W 365 S", measuring 572 x 183 x 170 mm and weighing 4.5 kg, nevertheless features all attributes of 'classic' ghetto blaster design (fig. 3). Two square speakers, with separate bass ports on the left and right hand side respectively, provide the necessary volume and depth of sound. Two cassette decks are positioned between the speakers. Featuring a green stripe, the left hand side deck is for playback, while, featuring a red stripe, the right hand side deck is used for recording. The decks are controlled by

Produkte werden nun in Abstimmung von Technologie und Kundenprofil entwickelt. So etwa das 1982 herausgebrachte „ICF 7600 A" von Sony (Kat.Nr. 142).[6] Ende der 1950er Jahre wurde mit dem Transistorradio das Radiohören auch außerhalb des heimischen Wohnzimmers möglich. Im Laufe der 1970er Jahre ist es möglich die verfeinerte Technik auf immer kleineren Raum zu reduziert, sodass mit praktischem Tragegriff ausgestattete Kompaktanlagen nun auch problemlos mit auf Reisen genommen werden können. Das japanische Modell „ICF-7600A" besticht durch sein Taschenbuchformat und geringes Gewicht.

Zu seiner ‚wahren Größe' und echter Berühmtheit findet der portable Radiorekorder mit integriertem Kassetten- oder CD-Deck jedoch im Laufe der 1980er Jahre durch die New Yorker HipHop-Szene. Als ‚Ghettoblaster' erobert er die Straßen und wird zum Freiheitssymbol einer Subkultur sowie Kultobjekt der Jugend.[7] Er liefert den richtigen Sound zur Breakdancewelle und taucht als wiederkehrendes Motiv in Graffiti und auf Plattencovern auf. 1985 veröffentlicht Virgin Interactive sogar ein nach dem Radiorekorder benanntes PC-Spiel. Sonys „CFS-W365S" von 1987 zählt mit seinen Maßen von 572 x 183 x 170 mm und einem Gewicht von 4,5 kg noch zu den kleineren Exemplaren der Gattung, doch erfüllt er alle Aspekte des ‚klassischen' Ghettoblaster-Designs (Abb. 3). Zwei quadratische Boxen links und rechts sorgen mit zwei separaten Bassports für ausreichend Lautstärke und Klangtiefe. Dazwischen finden zwei Kassettenlaufwerke Platz, von denen das linke, grafisch grün hervorgehoben, Abspiel- und das rechte, rot markiert, Aufnahmefunktion erfüllen.

3 / CFS-W365S, Sony Corp., Tokio (JP) 1987, Manuela Cirillo-Karpf
© Foto / Photo: Rheinisches Bildarchiv Köln / Cologne , Marion Mennicken.

a row of rocker switches below the clear cassette decks. Above the decks is a horizontal panel to select radio playback or cassette function, for tuning and for controlling sound quality and volume. On the typical simple black ghetto blaster casing the different functions are indicated only by unobtrusive colour accents. Used for tuning, the upper part of the panel is visually separated by a blue line from the cassette-recorder controls. Next to the three slide controls for the equaliser, the different frequency ranges are symbolised by white horizontal lines, which are visually framed by vertical ellipses in the same blue colour as the line separating the upper part of the control panel from the lower part. As the most important part of this portable stereo system, the two speakers are emphasised by chrome-effect bezels.

Graham Hinde's design for the "Roller 8007" by Philips (cat.no. 145) also works with colour accents to break up the device's functionally structured casing and to emphasise the circular and playful forms. Simple, but with a twist, the "Roller D 8007" was awarded the 1987 iF Design Award.

Its rational design, where the different technical components remain visible, has made the ghetto blaster the image-promoting symbol that is still referred to in contemporary music videos and song titles.[8]

However, within a decade, the ghetto blaster's popularity was waning. At the beginning of the 1990s, people returned to listening to the radio at home, using stationary stereo systems or, when not at home, the practical "Walkman". Heralding the advent of the Walkman, the German company Supertech launched its "Walky Box" (fig. 4). Available

Sie werden über eine Kipptastenreihe unterhalb der transparenten Kassettendecks bedient. Oberhalb dieser erstreckt sich ein horizontales Band zur Auswahl von Radio- oder Kassettenfunktion, Einstellung der Radiofrequenz sowie Ton- und Lautstärkeregelung. Auf dem für den Ghettoblaster typischen, schlicht schwarzen Kunststoffgehäuse sind die verschiedenen Funktionen lediglich durch zurückhaltende, farbliche Akzentuierungen markiert. Die obere Zone, die der Radioeinstellung vorbehalten ist, ist durch ein blaues Längsband vom Bereich des Kassetten-Rekorders abgesetzt. Um die drei Schiebeschalter für die Frequenzbestimmung sind die verschiedenen, mit weißen Querstrichen dargestellten Frequenzabschnitte durch senkrecht stehende Ellipsen im selben Blauton visuell hervorgehoben. Als wichtigstes Element der portablen Stereoanlage werden die Boxen durch silberne Umrahmungen betont.

Auch der Entwurf Graham Hindes für den Philips „Roller D8007" (Kat.Nr. 145) setzt auf farbliche Akzente, um das funktional gegliederte Gehäuse des Radiorekorders aufzulockern und die runden, spielerischen Formen zu unterstreichen. Schlicht, aber mit Pfiff, gewinnt das Modell 1987 den iF Design Award.

Ein nüchternes Erscheinungsbild, das die verschiedenen technischen Bestandteile erkennen lässt, hat den Ghettoblaster zum imageträchtigen Symbol avancieren lassen, welches bis heute in Musikvideos und Liedtiteln wachgerufen wird.[8]

Doch im Laufe des Jahrzehnts verblasst seine breite Popularität. Anfang der neunziger Jahre hört man Radio entweder wieder zuhause, nun aus der stationären Stereoanlage, oder unterwegs aus dem handlichen „Walkman". Auf dem Weg dorthin bringt die deutsche Firma Supertech ihre „Walky Box" (Abb. 4) heraus. In einer knalligen Farbskala erhältlich beherbergt das fast würfelförmige Gehäuse vorrangig einen Lautsprecher, der bis auf einen schmalen Rand die gesamte Front einnimmt. Teleskopantenne, Lautstärke- und Frequenzregler sitzen in der schwarz abgesetzten funktionalen Zone auf der Oberseite, die vom eckigen Tragegriff in derselben Farbkombination umfasst wird.

in various loud colours, an almost dice-shaped case mainly houses a speaker, which, apart from a narrow bezel, takes up the entire front. A telescopic aerial, as well as volume and tuning controls are positioned in a black functional zone on the device's top. This zone is framed by an angular handle featuring the same colour combination as the radio.

In postmodernism, individualism was spelt in capital letters. Accordingly, designers increasingly emerged from the anonymity of production workflows to enter the limelight as the creative minds behind the products. More than any other designer, Frenchman Philippe Starck has succeeded in becoming an international media star.[9] Made famous by his furniture and interior designs, he extends his creative energy to almost all areas of life.[10] He maintains a personal, eco-friendly design philosophy[11], advocating a considered use of materials without being spartan.[12]

Designed in 1990 for Thomson Multimedia, the portable "RT 201" radio appears, at first glance, to be a rather plain rectangular solid (cat.no. 154). The front of the small portable device combines the functions of speakers, display and controls on one single surface. The sound outlets come in the form of perforations that are evenly distributed over the entire front, thus simultaneously forming a decorative pattern. In its lower quarter, the surface features three depressions next to each other that accommodate spherical buttons to control an integrated alarm clock. In a display of subtle detailing, shallow grooves between the sound outlets create a visual connection between the buttons, labels and the company wordmark displayed below the buttons. Matching this design, a circular digital display located in the upper part of the rectangular body rises

4 / Walky Box, Supertech Deutschland GmbH, Nettetal (DE) 1990, Marion Brass
© Foto / Photo: Rheinisches Bildarchiv Köln / Cologne, Marion Mennicken.

Die Postmoderne schreibt Individualismus groß. Dem-entsprechend treten auch zunehmend die Designer als kreative Köpfe aus der anonymen Produktionskette heraus. Der Franzo-se Philippe Starck schafft es wie kein Zweiter zum internationa-len Medienstar.[9] Durch seine Möbelentwürfe und Inneneinrich-tungen bekannt geworden, breitet er seine kreative Energie auf nahezu alle Bereiche des menschlichen Lebens aus.[10] Er vertritt seine persönliche, umweltbewusste Gestaltungsphilosophie,[11] die den reduzierten Einsatz von Materialien propagiert, ohne dabei spartanisch daher zu kommen.[12]

Der tragbare Radioapparat „RT 201", der 1990 für Thomson Multimedia entsteht, erscheint zunächst als unauffälliger Qua-

from the surface in an organic curve. The dial-con-trolled tuning display is embedded along the vertical centre of the right-hand narrow side. Mirroring this arrangement, the on/off button, the band selection button, the alarm on/off button and the earphones socket are located on the left hand narrow side. The retractable aerial disappears into the top surface. Contrasting the geometric body, the collapsible curved supports at the back have a lightweight and playful appearance.

Like in his later radio designs for Alessi and Oregon Scientific (cat.no. 156 and 160), Starck suc-cessfully created a harmonious balance between organic and geometric forms. Using sophisticated detailing, he gives the device an individual per-sonality while remaining true to the principle of efficient functionality.

The 1994 "Radio 20" design by ISIS commu-nicates its task in a totally different way (cat.no. 152): the radio's form not only follows, but also states, its function. Forming the letters 'RADIO', the product's material shape represents its designation, thus play-fully going beyond the difference between designa-tor and designated object as explained by linguist Ferdinand de Saussure.[13] The elements for using the device are integrated in the most unobtrusive way possible: the speaker is fitted in the circular shape of the last letter, on top of which the controls are located (which are not visible when looked at from the front). The telescopic aerial can be collapsed and discreetly hidden away at the back.

The "RADIO Days" exhibition features different designs of the "Radio 20" model. While the design with the red lacquer finish lends the letter-object expressive force, another design uses a

der (Kat.Nr. 154). Die Front des handlichen Gerätes vereint Lautsprecherfunktion, Displayanzeige und Bedienungsfunktion auf einer Fläche. Die Schallöffnungen verteilen sich als gleichmäßiges Lochraster über die gesamte Schauseite und bilden somit zugleich eine dekorative Musterung. Im unteren Viertel vertieft sich die glatte Oberfläche in drei nebeneinander liegende, gleichmäßige Krater, die kugelförmige Tasten zur Einstellung der integrierten Uhr- und Weckfunktionen aufnehmen. Als dezentes Detail bilden flache Einkerbungen zwischen den umliegenden Lautsprecheröffnungen eine optische Verbindung von Taste, dazugehöriger Beschriftung und dem unterhalb dieser Zone befindlichen Firmenlogo. Als Pendant erhebt sich im oberen Abschnitt des Rechteckkörpers das kreisrunde digitale Displayfeld in einer organischen Wölbung aus der Oberfläche. An der rechten Schmalseite sind entlang der vertikalen Mittelachse Frequenzskala sowie dazugehöriger Drehregler eingelassen. Symmetrisch hierzu befinden sich auf der linken Längsseite Einschalttasten des Geräts, der Bandbreite und des Alarms sowie der Kopfhöreranschluss. Auf der Oberseite lässt sich die Teleskopantenne spurlos im Gehäuse versenken. Ausklappbare, geschwungene Stützen an der Rückseite wirken im Gegensatz zum geometrischen Korpus leicht und verspielt.

Wie in seinen späteren Radiodesigns für Alessi und Oregon Scientific (Kat.Nr. 156 und 160), gelingt es Starck organische und geometrische Formen in harmonische Ausgewogenheit zu bringen. Mithilfe raffinierter Details verleiht er dem Gerät eine persönliche, charakterhafte Note und bleibt dabei dennoch ökonomischer Funktionalität treu.

Auf ganz andere Weise thematisiert das Modell „Radio 20" der Firma ISIS von 1994 seine Aufgabe (Kat.Nr. 152). Seine Form orientiert sich nicht an der Funktion, sondern wird zu deren Verweis. Als Buchstabenkette „RADIO" stellt das Objekt seine materialgewordene Bezeichnung dar und überschreitet damit spielerisch die vom Linguisten Ferdinand de Saussure erläuterte Differenz zwischen bezeichnendem Begriff und bezeichnetem Objekt.[13] Für die Inbetriebnahme notwendige Ausstattungselemente werden möglichst unauffällig integriert: der Lautsprecher passt sich in die runde Öffnung des letzten Buchstabenkörpers ein, auf dem sich, bei Frontalansicht nicht zu sehen, die nötigen Bedienungsschalter befinden. Die Antenne ist ausziehbar und kann unauffällig im Rücken des Gerätes verstaut werden.

mirror facing for the letters, which, to some degree, dissolves the word-object's physical presence. This design represents the device's twofold nature as a medium of language. The constantly changing mirror image corresponds to the ephemeral sound signal, which, besides writing, is a form of the medium of language according to Niklas Luhmann's media theory.[14] Form and medium are two inseparably entwined prerequisites for communication, which, for Luhmann, is the basis of society.

The "Radio 20" receiver only allows one-way communication, but its mirror surface already integrates stimuli from its direct environment into its visible material gestalt.

Today, we are in a universe of formless media. With digital broadcasting, radio is just one of many options integrated into multi-functional devices. This has complex consequences for design. As radio stations can now also be experienced visually and in interactive ways in the form of apps on smartphones or as Internet platforms, both the producers of radio broadcasts and product designers are faced with totally new challenges.[15] People are no longer looking for acoustic information alone: visual components must also be provided. It is no longer the tactile casing of an object that has to convince users: rather it is the user-friendly and attractive design of a virtual interface because, with the World Wide Web, tuning is no longer dependent on locality, which is why radio stations are facing increased competitive pressure.[16]

In virtual space, the former 'listeners' are no longer passive receivers: they are active users who can listen to, and download, features and music via streaming or in the form of podcasts. And they can

In der Ausstellung „RADIO Zeit" sind verschiedene Ausführungen des „Radio 20" zu sehen. Während der rote Lack dem Wortkörper expressive Kraft verleiht, ist in einer anderen Version der Schriftzug mit einer spiegelnden Front versehen. Diese unterwandert die körperliche Präsenz des Wort-Objekts. Das Radiogerät verkörpert damit als Sinnbild seine doppelsinnige Realität als Sprachmedium. Das sich ständig aktualisierende Spiegelbild entspricht dem ephemeren Tonsignal, das neben der Schrift in der Medientheorie Niklas Luhmanns eine Form des Mediums Sprache ist.[14] Form und Medium bilden unlöslich miteinander verknüpfte Bedingungen für die Realisierung von Kommunikation, die für Luhmann die Grundlage der Gesellschaft bildet. Der Rundfunkempfänger „Radio 20" lässt Kommunikation nur einbahnig passieren, doch nimmt die spiegelnde Oberfläche bereits Reize der direkten Umwelt in die materielle, sichtbare Gestalt auf.

Heute befinden wir uns in einem Universum des formlosen Mediums. Der digitale Rundfunk macht Radio zu einer Option von vielen, integriert in multifunktionale Geräte. Dies bringt vielschichtige Folgen für das Design mit sich. Da Radiosender als App auf dem Smartphone oder Plattform im Internet nun auch visuell und interaktiv erfahrbar sind, werden neue Ansprüche an die Produzenten von Radiobeiträgen sowie an die Produktgestaltung gestellt.[15] Nicht mehr die rein akustische Information ist von Interesse, es werden visuelle Zusätze erforderlich. Nicht mehr das Gehäuse eines haptischen Objekts muss überzeugen, sondern das nutzerfreundliche und attraktive Design einer virtuellen Oberfläche; denn im World Wide Web gerät der einzelne Radiosender ohne Einschränkung lokal empfangbarer Frequenzbereiche unter erhöhten Konkurrenzdruck.[16]

Der ehemalige Radiohörer ist im virtuellen Raum kein passiver Empfänger, sondern aktiver Nutzer, der sich Beiträge und Musiktitel per Streaming oder in Form von Podcasts in beliebiger Reihenfolge überall und unabhängig vom durch den Sender vorgegebenen Programm anhören und abspeichern kann. Darüber hinaus kann er eigene Beiträge entwerfen und diese über das Internet zur Verfügung stellen. Löste der Spiegel die visuelle Objektgrenze des „Radio 20" auf, wird das formlose Radio zur durchlässigen Membran.

do so in whichever order they want, independently of programme structures defined by radio stations. Moreover, users can now create their own contributions and publish them on the Internet. While the mirror dissolved the visual object borders of the "Radio 20", the formless radio becomes a permeable membrane.

In response to these developments, current radio design is experiencing a wave of nostalgia, with 'retro style' dominating the market. Households are full of colourful devices in many different shapes, with new technology 'dressed' in the classic designs of old valve radios and portable radios.

1   This motto captures the gist of Venturi's statements on Mies van der Rohe's principles. Although often cited in design history, there is no proof that Venturi used this exact wording. See Venturi, vol. 1, 1966: 24-25.

2   Berents 2011: 170.

3   Sparke 2002: 238-239.

4   Andy Warhol as cited in: Celant 1998-1999, n.pag.

5   See Sparke 2002: 180-181.

6   The ICF-7600A is the direct successor to the 1978 Sony "ICF-7600", where the large and clearly laid out frequency dial is used for the first time in a small device. See Bösch 2005.

7   See Hengstenberg 2009.

8   For example in the single Ghetto Blastah by Norwegian DJ Savant.

9   Starck is one of the few living designers to whom the Centre Pompidou has dedicated a comprehensive retrospective. See Guillaume 2003; on Starck's website there is also an impressive list of international publications that have devoted special editions to him: http://www.starck.com/fr/presse/revue_de_presse/speciaux_starck/ [25.9.2015].

10 Starck advocates the idea, related to modernist utopian thinking, that appealing product design can persuade people to change their habits, thus leading to a happier life. Around 1998, for example, he designed packaging for organic food. See Starck 2003: 71-73.

11 See Berents 2011: 204-208.

12 Colin 1988: 24.

In Reaktion auf diese Entwicklung macht sich im aktuellen Radiodesign Nostalgie breit. Der Retro-Style beherrscht den Markt. Neue Technik im Gewand alter Röhrenapparate und bunter Kofferradios bevölkert farbenfroh und formenreich die Haushalte.

1   Dieses Motto entspricht dem konzentrierten Sinngehalt Venturis Aussagen und Stellungnahmen über die Grundsätze von Mies van der Rohe. Obwohl in der Designgeschichte daher vielfach zitiert, kann ihm das Bonmot jedoch nicht wortwörtlich nachgewiesen werden. Vgl. Venturi, Bd. 1, 1966, S. 24-25.

2   Berents, 2011, S. 170.

3   Sparke, 2002, S. 238-239.

4   Andy Warhol zitiert nach: Celant, 1998-1999, o.S.

5   Vgl. Sparke, 2002, S. 180-181.

6   Es handelt sich hierbei um den direkten Nachfolger des Sony „ICF-7600" aus dem Jahr 1978, bei dem zum ersten Mal die großformatige, übersichtliche Frequenzskala im handlichen Format eingesetzt wird. Vgl. Bösch, 2005.

7   Vgl. Hengstenberg, 2009.

8   So etwa 2012 in der Single „Ghetto Blastah" des norwegischen DJs Savant.

9   Als einem von wenigen Vertretern seines Metiers wurde Starck zu Lebzeiten eine großangelegte Retrospektive im Centre Pompidou gewidmet. Vgl. Guillaume, 2003; zudem kann auf der Homepage des Designers eine beeindruckende Liste ihm gewidmeter Sonderausgaben internationaler Presseveröffentlichungen betrachtet werden: http://www.starck.com/fr/presse/revue_de_presse/speciaux_starck/ (25.9.2015).

10 Starck vertritt die modernistischen Utopien nahestehende Idee, dass ansprechendes Produktdesign Überzeugungsarbeit leisten kann, die durch Veränderung menschlicher Gewohnheiten zu einem glücklicheren Leben führen kann. So entwirft er etwa 1998 eine Verpackungsreihe für Bio-Nahrungsmittel. Vgl. Starck, 2003, S. S. 71-73.

11 Vgl. Berents, 2011, S. 204-208.

12 Colin, 1988, S. 24.

13 Heringer, 2013.

14 Der flüchtige Laut ist nach der Medientheorie Niklas Luhmanns ebenso wie die Schrift nur eine temporär sich vollziehende Form des Mediums Sprache, das gemessen an diesen Ausformungen dauerhaft ist. Vgl. Luhmann 1998.

15 Einen vielseitig reflektierter Ausblick auf die zukünftige Gestaltung von Radioprogrammen und die neu entstehenden Ansprüche an das Medium Radio im Kontext der Digitalisierung findet man in: Koch/Glaser, 2005.

16 Vgl. Zierl, 2011, S. 241.

13 Heringer 2013.

14 According to Niklas Luhmann's media theory, like writing, ephemeral sound is a fleetingly occurring form of the medium of language. Compared to these forms, language is a permanent medium. See Luhmann 1998.

15 A comprehensive analysis of what the future may hold with regard to the design of radio programmes and to the new demands placed upon the medium of radio in the context of digitisation can be found in: Koch/Glaser 2005.

16 See Zierl 2011: 241.

Katalog

Collections
Catalogue

Werbeschild / Advertising sign
Stromberg-Carlson
Rochester, New York (US)
1925-40
Metall; Glas
        Metal; glass
28 x 50 x 14 cm
Inv.Nr. K 1535 W

Radiola 17
RCA
Holzgehäuse / wooden casing: Stout-Smith Trust
Salem, Indiana (US)
1927-28
Holz; Messing
        Wood; brass
23 x 70 x 21 cm
Inv.Nr. K 1364 W

4

Lautsprecher / Speaker
RCA
New York City, New York (US)
1928-29
Aluminiumguss (?), lackiert
        Casted aluminium (?), lacquered
21,5 x 29,5 x 13,5 cm
Inv.Nr. K 1356 W

Lautsprecher / Speaker, 2007 „Bratpfanne"
N. V. Philips' Gloeilampenfabrieken
Eindhoven (NL)
1928-31
Press-Phenolharz (Philite)
        Pressed phenolic resin (Philite)
43 x 40 x 18 cm
48 x 44,5 x 20 cm
Inv.Nr. K 1351 W
Inv.Nr. K 1352 W

Volksempfänger VE 301 Wn
Walter Maria Kersting
Mende & Co.
Dresden (DE)
1933-37
Press-Phenolharz (Bakelit), Textil
        Pressed phenolic resin (Bakelit); textile
39 x 28 x 17 cm
Inv.Nr. K 793

Volksempfänger VE 301
Walter Maria Kersting
Mende & Co.
Dresden (DE)
1933-38
Press-Phenolharz (Bakelit), Textil
        Pressed phenolic resin (Bakelit); textile
39,5 x 28 x 19,5 cm
Inv.Nr. K 1324 W

7

Eine wahre Ikone der frühen Kunststoffgehäuse stellt das Ekco „AD 65" dar. Es ist das erste Radio in Kreisform überhaupt. Der Architekt Wells Coates reichte den Entwurf 1932 bei dem britischen Hersteller E.K. Cole & Son – kurz Ekco – ein, der einen Wettbewerb für Radiogehäuse in Press-Phenolharz ausgeschrieben hatte. Ekco war eines der frühesten Unternehmen, das seine Produktion auf Kunststoff umstellte. Diese Aufgabe kam dem kanadisch-stämmigen Coates entgegen, der sehr an kostenoptimierten Lösungen für Industrieprodukte interessiert war. Seine Maxime lautete „purpose related to purse", also dass Zweck und Ziel einer Gestaltung auch vom Geldbeutel abhängen. Coates gewann den Wettbewerb und verwies mit dem einzigen runden Design die namhaften Konkurrenten auf die Plätze. Ein Grund des Erfolgs lag sicherlich darin, mit dem Entwurf die Qualität des Materials Kunststoff adäquat zum Ausdruck bringen zu können. Das übliche (eckige) Hoch- oder Querformat leitete sich von den Holzkisten ab, in die man die Technik häufig verpackte. Das „AD 65" ging 1934 in Produktion und wurde mit seiner eigenständigen Form und dennoch klaren Anlehnungen an Art Déco-Ästhetik – dunkles Gehäuse mit Chromelementen – ein Erfolgsschlager. Das Modell gab es zunächst in Braun und Schwarz. Ironischerweise war der braune Korpus bei den Käufern beliebter, da die Farbe an Nussholz erinnerte. Das schwarze Modell wirkte demgegenüber ‚industrieller'.

Ekco AD 65
Wells Coates, 1932
Erik Kirkham Cole Limited
Southend-On-Sea (GB)
1934
Press-Phenolharz (Bakelit)
        Pressed phenolic resin (Bakelit)
39,5 x 38 x 21 cm
Inv.Nr. K 1403 W

The Ekco "AD 65" is a true icon of early plastic casing design. It was the first-ever circular radio. Architect Wells Coates entered the design in 1932 to a competition by British manufacturer E.K. Cole & Son, or Ekco for short, who were looking for radio casing designs using pressed phenolic resin. Ekco was one of the first companies switching their production to plastic. Coates was the right man for the task as he was particularly interested in cost-efficient solutions for mass produced products. For him, the purpose and objective of design was also dependent on the money that the target users were able to spend, captured in his maxim "purpose related to purse". With the only circular design, Coates won the competition, leaving his renowned competitors upstaged. One of the reasons for this success will certainly have been the fact that his design perfectly showcased the material's unique qualities. The conventional vertical or horizontal rectangular shapes were derived from the wooden boxes in which the technology used to be housed. The "AD 65" was produced and, with its unique form that nonetheless featured clear references to art déco aesthetics (dark colours and chromed details), became a top seller. The design was initially available in brown and black. Ironically, the brown casing proved to be the most popular as the colour was reminiscent of walnut while the black design had a stronger industrial feel.   RB

Ekco AC 85 Superhet
Wells Coates
Erik Kirkham Cole Limited
Southend-On-Sea (GB)
1934
Press-Phenolharz (Bakelit)
　　　Pressed phenolic resin (Bakelit)
31,8 x 53,3 x 24,1 cm
Inv.Nr. K 1292 W

66 Skyscraper
Air-King Products Company
Harold L. van Doren, John Gordon Rideout
Brooklyn, New York (US)
1935
Harnstoffharz (Plaskon)
　　　Urea resin (Plaskon)
30 x 23 x 19 cm
Inv.Nr. K 1251 W

10

Sparton 1186 Nocturne
Walter Dorwin Teague
Sparks-Withington Company
Jackson, Michigan (US)
1935-37
Glas; Metall; Holz
     Glass; metal; wood
117 x 110 x 38 cm
Inv.Nr. K 1375 W

Als im September 1935 die „Annual National Electrical &
Radio Exposition" in New York ihre Pforten öffnete, wartete die
Sparks-Withington Company (Sparton) mit einer Sensation auf:
Sie enthüllte feierlich die „1186 Nocturne", ein Bodenradio aus
mitternachtsblauem Spiegelglas (Durchmesser 110 cm) mit sati-
nierten Chromelementen. Walter Dorwin Teague, der die verspie-
gelte Schönheit geschaffen hatte, verfolgte bei seinen Entwürfen
das Ziel, traditionelle Ästhetik – wie beispielsweise die klassische
Proportionslehre – mit innovativer Formensprache zu verbinden.
Der horizontale Durchmesser der „Nocturne" ist in drei gleiche
Abschnitte geteilt, es ergeben sich in der Vertikalen drei gleiche
‚Streifen'. Der vertikale Durchmesser wiederum besitzt ein beson-
deres Maß, nach dem sich Lautsprechereinheit im unteren Bereich sowie Skala, Bedienknöpfe und Magisches Auge im oberen
Bereich organisieren. Teague teilte die Strecke zweimal durch die
Zahl Phi (1,618...), so dass sich die Proportionen nach den Gesetzmäßigkeiten des Goldenen Schnitts richten. Das Ende der Strecke a – also der erste Goldene Schnitt – befindet sich von unten
aus gesehen an der untersten Chromleiste der Zone mit den Bedienelementen. Durch erneute Teilung der verbliebenen Strecke
b liegt der zweite Goldene Schnitt exakt dort, wo das Magisches
Auge ansetzt. – Die „Nocturne" gilt als Inbegriff und Höhepunkt
der amerikanischen Art Déco-Radios.

When, in September 1935, the Annual National Electrical &
Radio Exposition opened its doors in New York, the Sparks-
Whittington Company (Sparton) had a sensation in store: in a
coup de théatre, Sparton unveiled its "1186 Nocturne", a large
floor-standing radio (110 cm diameter) encased in midnight-blue
mirror glass with satin-chrome details. Walter Dorwin Teague,
the creator of the mirror-glass beauty, wanted to combine traditional aesthetics – such as classical laws of proportion – with an
innovative design language. The Nocturne's horizontal diameter
is divided into three identically sized sections, resulting in three
identical ‘stripes' in the vertical plane. The vertical diameter, on
the other hand, features special dimensions into which the
speaker in the lower part, as well as the dial, controls and the
magic eye in the upper part are organised. Teague divided the
line twice by the number phi (1.618...) so that the proportions followed the laws of the golden section. Viewed from the bottom,
the endpoint of line a – i.e. of the first golden section – is located at the lowest chromed bezel of the zone with the control elements. By again dividing the remaining line b, the second golden
section is positioned exactly at the point where the magical eye
starts. – The Nocturne is seen as the epitome and climax of
American art déco radios.   RB

Sparton 566 Bluebird
Walter Dorwin Teague
Sparks-Withington Company
Jackson, Michigan (US)
1935-36
Holz; Glas; Metall, verchromt
        Wood; glass; chrome-plated metal
36,5 x 37 x 18,5 cm
Inv.Nr. K 1376 W

Silvertone 4500 Election
John R. Morgan
Sears, Roebuck & Co.
Chicago, Illinois (US)
1936
Press-Phenolharz (Bakelit)
        Pressed phenolic resin (Bakelit)
18 x 24,3 x 14,5 cm
Inv.Nr. K 1370 W

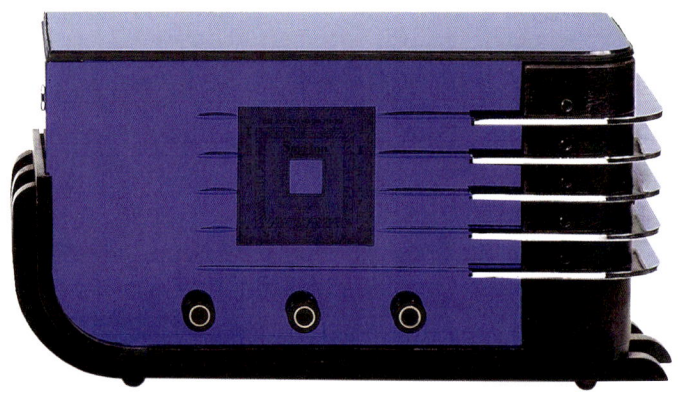

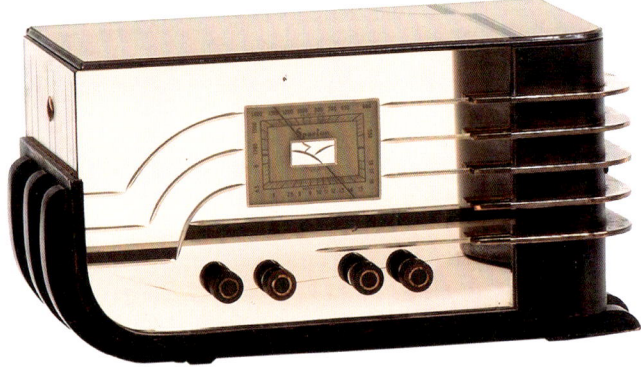

13

Sparton 557 Sled
Walter Dorwin Teague
Sparks-Withington Company
Jackson, Michigan (US)
1936-37
Glas; Holz; Metall
        Glass; wood, metal
23 x 45,5 x 21 cm
Inv.Nr. K 1380 W

Sparton 558 C Sled
Walter Dorwin Teague
Sparks-Withington Company
Jackson, Michigan (US)
1937-39
Holz; Glas; Press-Phenolharz (Bakelit); Kupfer
        Wood; glass; pressed phenolic resin (Bakelit);
        copper
22,5 x 45,5 x 21 cm
Inv.Nr. K 1379 W

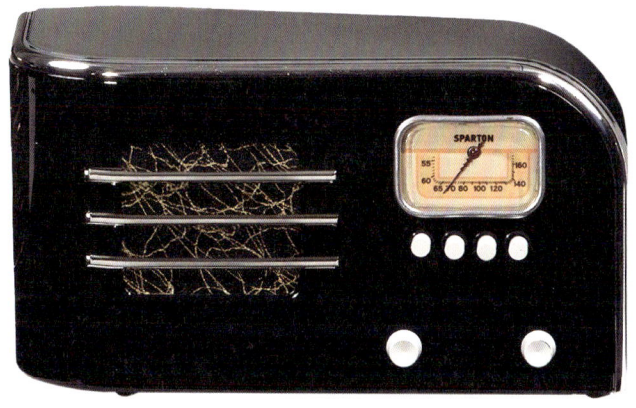

Sparton 5518 Selectronne
Walter Dorwin Teague
Sparks-Withington Company
Jackson, Michigan (US)
1938
Metall
        Metal
21 x 38 x 18,5 cm
Inv.Nr. K 1378 W

6 D 311
Zenith Radio Corp.
Chicago, Illinois (US)
1938
Press-Phenolharz (Bakelit)
        Pressed phenolic resin (Bakelit)
16,7 x 27 x 19 cm
Inv.Nr. K 1392 W

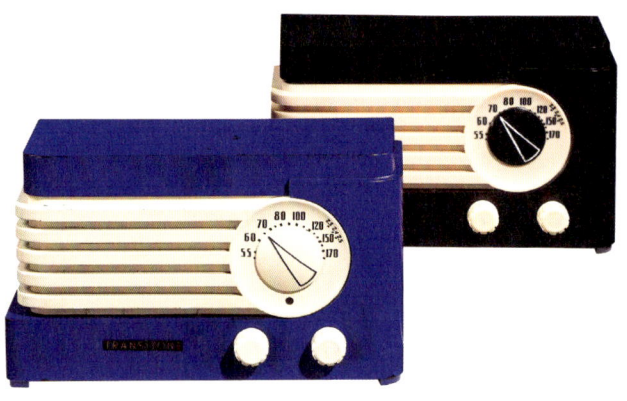

A 5 Dwarf
J. Samson Spencer
Halson Radio Manufacturing Co.
New York City, New York (US)
1938
Guss-Phenolharz (Catalin)
    Cast phenolic resin (Catalin)
12,5 x 18,5 x 13 cm
Inv.Nr. K 1311 W

Transitone TP 10
Philco
Philadelphia, Pennsylvania (US)
1938-39
Press-Phenolharz (Bakelit), lackiert
    Pressed phenolic resin (Bakelit), lacquered
13,5 x 23 x 12 cm
Inv.Nr. K 1347 W (blau / blue)
Inv.Nr. K 1348 W (schwarz / black)

Silvertone 6110 Rocket, „Turbine"
Clarence Karstadt
Sears, Roebuck & Co.
Chicago, Illinois (US)
1938-39
Press-Phenolharz (Bakelit)
          Pressed phenolic resin (Bakelit)
17 x 30 x 16,5 cm
Inv.Nr. K 1369 W

Volksempfänger VE 301 Dyn
Radio A.G. D.S. Loewe
Berlin-Steglitz (DE)
1938-41
Press-Phenolharz (Bakelit)
          Pressed phenolic resin (Bakelit)
31,5 x 27,5 x 21 cm
Inv.Nr. K 1322 W

Deutscher Kleinempfänger DKE 38
„Goebbels Schnauze"
Mende & Co.
Dresden (DE)
1938-44
Press-Phenolharz (Bakelit)
        Pressed phenolic resin (Bakelit)
24 x 24 x 13,5 cm
Inv.Nr. K 1323 W

TW 49
Sonora Radio & Television Corp.
Chicago, Illinois (US)
1939-40
Press-Phenolharz (Bakelit)
        Pressed phenolic resin (Bakelit)
17 x 28 x 16 cm
Inv.Nr. K 1371 W

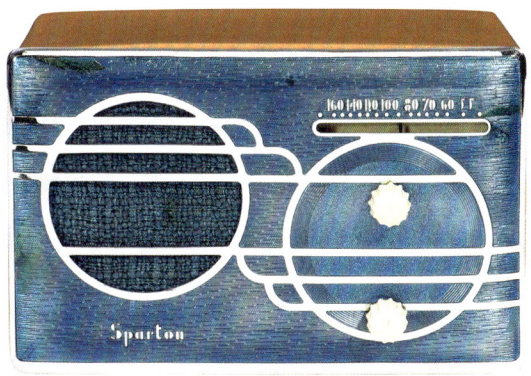

Sparton 500 C Cloisonné
Walter Dorwin Teague
Sparks-Withington Company
Jackson, Michigan (US)
1939-40
Guss-Phenolharz (Catalin); Metall, verchromt;
Emaille
        Cast phenolic resin (Catalin); chrome-plated
        metal; enamel
13 x 20 x 11 cm
Inv.Nr. K 1377 W

Motorola 51 X-15; -16 „S-Grill"
Galvin Manufacturing Corp.
Chicago, Illinois (US)
1939-42
Guss-Phenolharz (Catalin)
        Cast phenolic resin (Catalin)
17,5 x 24,5 x 16 cm
Inv.Nr. K 1331 W (gelb / yellow, -16)
Inv.Nr. K 1332 W (schwarz / black, -15)

400-1; -2; -3 Patriot
Norman Bel Geddes
Emerson Radio and Phonograph Corp.
New York City, New York (US)
1940
Guss-Phenolharz (Catalin); Celluloseacetat (Tenite);
Harnstoffharz
          Cast phenolic resin (Catalin); cellulose acetate
          (Tenite); urea resin
20,5 x 27,5 x 16 cm
Inv.Nr. K 1297 W (blau / blue, -1)
Inv.Nr. K 1296 W (weiß / white, -2)
Inv.Nr. K 1295 W (rot / red, -3)

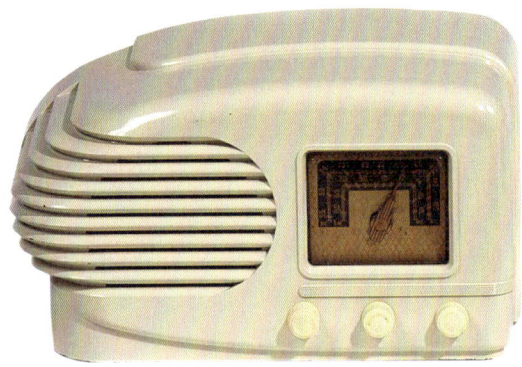

400 Aristocrat
Norman Bel Geddes
Emerson Radio and Phonograph Corp.
New York City, New York (US)
1940
Guss-Phenolharz (Catalin)
        Cast phenolic resin (Catalin)
20,5 x 27,5 x 16 cm
Inv.Nr. K 1298 W

Goldentone 1 W 10
Griffin-Smith Manufacturing Co.
Los Angeles, Kalifornien (US)
1940-41
Kunststoff, lackiert
        Plastic, lacquered
18,5 x 33,5 x 19 cm
Inv.Nr. K 1244 W

200; 1000 Bullet, Streamliner
FADA Radio & Electric Co.
Long Island, New York (US)
1940-41 (200); 1945-46 (1000)
Guss-Phenolharz (Catalin)
         Cast phenolic resin (Catalin)
16,5 x 26 x 15,5 cm
Inv.Nr. K 1300 W (weiß / white, 1000)
Inv.Nr. K 1301 W
(gelb, ohne Abb. / yellow, without fig., 1000)
Inv.Nr. K 1302 W (blau / blue, 200)

Das FADA „Bullet" oder „Streamliner" gehört zu den be-kanntesten amerikanischen Catalin-Radios. Die beiden Namen, die für alle Modellnummern verwendet wurden, beziehen sich auf die dynamische Form des Gehäuses. Das Streamline-Design galt in den USA von ca 1930 bis 1955 als Inbegriff des Fort-schritts und der Moderne. Die Stromlinie erfasste in diesem Zeit-raum quasi alle Industrieprodukte von der Lokomotive bis zum Bügeleisen. Das Gehäuse wurde 1940 entworfen und mit der Mo-dellnummer 115 auf den Markt gebracht. Die möglichen Farben und Farbkombinationen lauteten Alabaster, Lapis Lazuli, Onyx und Rubin – was den Anspruch auf den Vergleich mit Edelstei-nen verdeutlicht. Die kostbare Anmutung setzt sich bei allen Modellen auch in der Skala fort: Die an eine Flugzeug-Armatur angelehnte Skala ist mit goldfarbener, geprägter Metallfolie abge-setzt. Allen Radios sind auch die Details der einfachen Handha-bung gemein. Der „Streamliner" verfügt über einen absenkbaren Griff und eine im Gehäuse verbaute Antenne, die FA-DA-SCOOP. Der dazugehörige Slogan lautete „Plug in, Play", also „Einstecken, Spielen". Obwohl der Apparat nur mit Netzbetrieb funktionierte, wurde so das Mitnehmen des Radios von einem Ort zum ande-ren attraktiv beworben. Der Erfolg der 115er bis 200er Serie veran-lasste FADA nach dem Produktionsstopp von 1941-45 das Modell als 1000er mit nur leichten Modifikationen wieder aufzunehmen. Die frühen Modellreihen verfügen über glatte, runde Drehregler, das 1000er hat Knöpfe mit seitlicher Riffelung.

The FADA "Bullet" or "Streamliner" is among the most well-known American Catalin radios. Used for all models in the range, the two names refer to the casing's dynamic form. In the USA, from about 1930 to 1955, the streamline design was regard-ed as the epitome of progress and modernism. During this peri-od, the streamline could be found in virtually all industrial prod-ucts, from locomotives to irons. The casing was designed in 1940 and launched with the model number 115. Indicating an intended comparison with precious stones, the available colours and col-our combinations were: alabaster, lapis lazuli, onyx and ruby. In all models, the notion of luxury is also expressed in the design of the dial: reminiscent of an aircraft dashboard, the dial is finished with gold-coloured embossed metallic foil. The different mod-els also share the same detailing for ease-of-use. The Streamliner features a retractable handle and an aerial that can be embedded in the casing – the FA-DA-SCOOP – and came with the slogan 'Plug in, Play'. Although the device could only be mains-operat-ed, the slogan nicely captured the fact that the radio could easily be taken from one room to another. After production was discon-tinued from 1941-45, due to the success of the 115 to 200 ranges, FADA started to use this design again with only a few modifica-tions in its 1000 range. The earlier models feature smooth rotary controls while in the 1000 range the controls are notched. RB

Silvertone 3451 Commentator
John R. Morgan
Sears, Roebuck & Co.
Chicago, Illinois (US)
1940-41
Kunststoff
     Plastic
15,3 x 27 x 15 cm
Inv.Nr. K 1368 W

L 570
General Electric Co.
Ontario, Kalifornien (US)
1942
Guss-Phenolharz (Catalin)
     Cast phenolic resin (Catalin)
19 x 24,3 x 17,3 cm
Inv.Nr. K 1613 W

5 G 003 ZZ Holiday
Robert D. Budlong
Zenith Radio Corp.
Chicago, Illinois (US)
1945-47
Kunststoff
        Plastic
28,4 x 33 x 15,3 cm
Inv.Nr. K 1397 W

5681
International Detrola Corp.
Detroit, Michigan (US)
1945-47
Metall, verchromt
        Chrome-plated metal
22,5 x 29 x 17,7
Inv.Nr. K 1290 W

6 AU 1 Commander
Garod Radio Corp.
New York City, New York (US)
1945
Guss-Phenolharz (Catalin)
        Cast phenolic resin (Catalin)
18,5 x 29,3 x 18,5 cm
Inv.Nr. K 1307 W (weiß / white)
Inv.Nr. K 1308 W
(gelb, ohne Abb. / yellow, without fig.)
Inv.Nr. K 1309 W (rot / red)

H 125 Little Jewel, „Refrigerator"
Westinghouse Electric Corp.
Sunbury, Pennsylvania (US)
1945-46
Press-Phenolharz (Bakelit), lackiert; Metall
        Pressed phenolic resin (Bakelit), lacquered;
        metal
24 x 15,2 x 16 cm
Inv.Nr. K 1389 W a (grün, matt / green, matted)
Inv.Nr. K 1389 W b (grün, glänzend / green, glossy)
Inv.Nr. K 1389 W c (schwarz / black)

66 X-7; -8; -9 „Tuna Boat"
RCA-Victor
New York City, New York (US)
1946
Guss-Phenolharz (Catalin)
        Cast phenolic resin (Catalin)
23,8 x 38,5 x 19 cm
Inv.Nr. K 1361 W
(schwarz, ohne Abb. / black, without fig., -7)
Inv.Nr. K 1362 W
(dunkelbraun, ohne Abb. / dark brown,
without fig., -9)
Inv.Nr. K 1363 W a
(dunkelrot, ohne Abb. / dark red, without fig., -8)
Inv.Nr. K 1363 W b (dunkelrot / dark red, -8)

Die RCA Victor Modellreihe „66 X" besticht mit ihrem außergewöhnlichen Äußeren: Es handelt sich um das größte je gefertigte Guss-Phenolharz-Gehäuse, dessen Ecken und Kanten für diese Technik ungewöhnlich sind. Aussparungen gibt es nur für die waagerechten Lautsprecherschlitze und die oben quer eingelassene Senderskala aus konvex gekurvtem Glas. Dies begünstigt vergleichsweise große, ruhige Flächen, die den Glanz der polierten Oberfläche betont. Zwar gab es die Serie auch aus einfarbigem Material (beispielsweise Schwarz wie bei „K 1361 W"), begehrter waren aber die marmorierten Gehäuse. Die beiden Modelle (K 1363 W a + b) in Ochsenblut-Rot weisen an der Front sehr feine Farbwirbel auf. An den Wangen und der Oberseite lösen sich diese in breitere Strukturen auf, die wie geschnittener Achat anmuten. Das „K 1362 W" zeigt mit einer Marmorierung auf Basis von Schokoladenbraun eine ausgesprochen seltene Farbvariante. Allen Gehäusen gemeinsam ist die glänzende Oberfläche, die von den etwas matteren, opak durchgefärbten Drehreglern kontrastiert wird. Die Knöpfe bestehen augenscheinlich auch aus einem anderen Material, vermutlich Press-Phenolharz. Der 1919 gegründeten Radio Corporation of America (RCA) gelang 1929 mit dem Kauf der Victor Talking Machine Company der große Coup: Der damals größte Schallplattenkonzern der Welt war berühmt durch seine Werbebotschaft „His Masters Voice", bei der ein Hund in einen Grammophonlautsprecher horcht.

The "66X" model range from RCA Victor stands out by virtue of its extraordinary casing: the largest moulded phenolic resin casing ever produced that features edges and angles which are unusual for this type of technique. Recesses are provided only for the horizontal speaker grilles and for the convex glass dial set in the casing's top part. This design allows relatively large, uninterrupted surfaces that showcase the lustre of the casing's polished finish. Although the range was also available in solid colours (for example, in black, as in the "K 1361 W"), the marbled casings proved to be more popular. The two models in oxblood red feature very fine colour swirls at the front. At the sides and on the top, the delicate swirls gradually form into a stronger pattern that looks like cut agate. With its marble pattern based on chocolate brown, the "K 1362 W" features a very rare colour variation. All casings share the same polished finish, contrasted by slightly matt, opaque rotary controls. The controls seem to be made from a different material, presumably from pressed phenolic resin. With its 1929 acquisition of the Victor Talking Machine Company, the Radio Corporation of America (RCA, established in 1919) landed a big coup: the then-largest record company in the world was famous for its advertising slogan 'His Master's Voice' paired with a logo showing a dog listening to a gramophone speaker.   RB

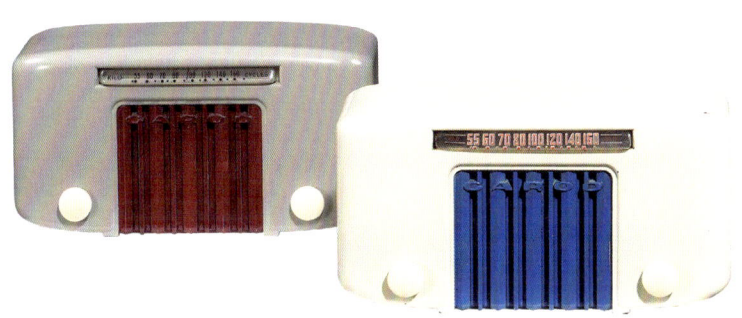

5 A 2 The Commodore
Garod Radio Corp.
New York City, New York (US)
1946
Press-Phenolharz (Bakelit), lackiert
　　　Pressed phenolic resin (Bakelit), lacquered
14,5 x 27 x 12,5 cm
Inv.Nr. K 1304 W (weiß / white)
Inv.Nr. K 1305 W (grau / grey)
Inv.Nr. K 1306 W
(grün, ohne Abb. / green, without fig.)

416 Frog Eyes
Barton T. Setchell
Setchell Carlson Inc.
St. Paul, Minnesota (US)
1946
Harnstoffharz (Plaskon)
　　　Urea resin (Plaskon)
17 x 25,5 x 16,5 cm
Inv.Nr. K 1367 W

39

6 D 111 „Rabbit", „Streamliner"
Belmont Radio Corp.
Chicago, Illinois (US)
1946
Kunststoff
        Plastic
19 x 32 x 17 cm
Inv.Nr. K 1259 W

5 D 128 Serie A
Belmont Radio Corp.
Chicago, Illinois (US)
1946
Kunststoff
        Plastic
17,7 x 30,5 x 18 cm
Inv.Nr. K 1257 W

Motorola 55 X 11 A
Galvin Manufacturing Corp.
Chicago, Illinois (US)
1946
Kunststoff
  Plastic
17 x 28,5 x 17 cm
Inv.Nr. K 1327 W

526 C
Frank Glover
Bendix Radio
Baltimore, Maryland (US)
1946
Guss-Phenolharz (Catalin)
  Cast phenolic resin (Catalin)
27,5 x 17,5 x 17 cm
Inv.Nr. K 1261 W

6 D 030 Z
Charles und / and Ray Eames
Evans für Zenith Radio Corp.
Chicago, Illinois (US)
1946
Formholz
        Plywood
20 x 35 x 16,5 cm
Inv.Nr. K 1739

571 C
Raymond Loewy
Aria International Detrola Corp.
Detroit, Michigan (US)
1946
Kunststoff; Metall
        Plastic; metal
18,5 x 30 x 18 cm
Inv.Nr. K 1252 W

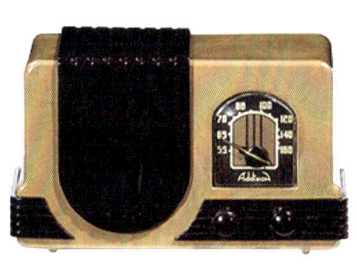
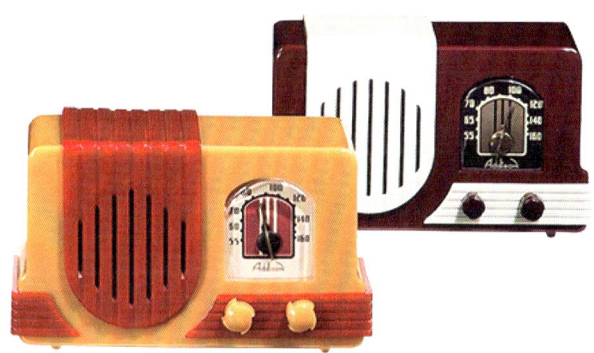

44

L 2 F; 52; 52 A; B 2 B; L 2 G
Addison
Toronto (CA)
1946-48
Guss-Phenolharz (Catalin)
      Cast phenolic resin (Catalin)
15,5 x 26,5 x 14,5 cm
Inv.Nr. K 1245 W (gelb / yellow, L 2 F)
Inv.Nr. K 1246 W (weiß / white, 52)
Inv.Nr. K 1247 W
(schwarz, ohne Abb. / black, without fig., 52)
Inv.Nr. K 1248 W (rot / red, 52 A)
Inv.Nr. K 1249 W (blau / blue, B 2 B)
Inv.Nr. K 1250 W (braun / brown, L 2 G)

B Lucite Radio
Cyarts Plastics Inc.
New York City, New York (US)
1946-47
Polymethylmethacrylat (Lucite); Kunststoff (Plexon)
        Polymethyl methacrylate (Lucite); plastic
        (Plexon)
18,6 x 36 x 16 cm
Inv.Nr. K 1619 W

444 A
Arvin
Columbus, Indiana (US)
1946-47
Metall, lackiert
        Metal, lacquered
13 x 17 x 12,3 cm
Inv.Nr. K 1253 W

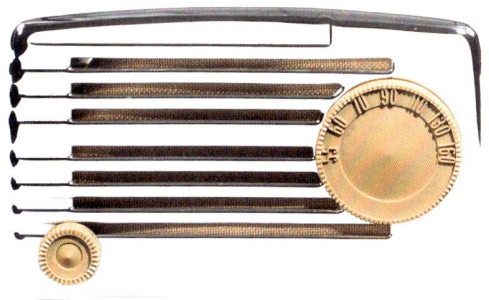

845 Cloud
FADA Radio & Electric Co.
Long Island, New York (US)
1946-50
Guss-Phenolharz (Catalin)
        Cast phenolic resin (Catalin)
17 x 26,7 x 15 cm
Inv.Nr. K 1303 W a (weiß / white)
Inv.Nr. K 1303 W b (beige / beige)

Ohne Modellbezeichnung / without model name
US
1946-55
Metall, verchromt; Kunststoff
        Chrome-plated metal; plastic
13,7 x 21 x 11 cm
Inv.Nr. K 1402 W

Crosley 56 TD Duette
Division AVCO Manufacturing Corp.
Cincinnati, Ohio (US)
1947
Press-Phenolharz (Bakelit), lackiert
      Pressed phenolic resin (Bakelit)
21 x 29 x 12 cm
Inv.Nr. K 1278 W (weiß / white)
Inv.Nr. K 1279W (rot / red)

A 100 „Toaster"
A. F. Thwaites
Murphy Radio Ltd.
Welwyn Garden City (GB)
1947
Press-Phenolharz (Bakelit)
      Pressed phenolic resin (Bakelit)
19 x 23,2 x 11 cm
Inv.Nr. K 1343 W

35 L 6
Belmont Radio Corp.
Chicago, Illinois (US)
1947
Kunststoff
        Plastic
19,5 x 30,5 x 19,5 cm
Inv.Nr. K 1258 W

Shure Mikrophone 556 B Unidyne
Ben Bauer
Shure Brothers Company
Chicago, Illinois (US)
1947
Aluminium; Metall; Kunststoff
        Aluminium; metal; plastic
21,5 x 8 x 9 cm
Inv.Nr. K 1732 W

114; 115
Frank Glover
Bendix Radio
Baltimore, Maryland (US)
1948
Kunststoff
        Plastic
18,5 x 28 x 17,5 cm
Inv.Nr. K 1260 W (hellrot / light red, 114)
Inv.Nr. K 1617 W (beige / beige, 115)

438190 Race Track
Coronado Manufacturing Corp.
Chicago, Illinois (US)
1947
Press-Phenolharz (Bakelit), lackiert; Celluloseacetat
(Tenite)
        Pressed phenolic resin (Bakelit), lacquered;
        cellulose acetate (Tenite)
19,3 x 24 x 16,8 cm
Inv.Nr. K 1273 W

WAU 243
Sonora Radio & Television Corp.
Chicago, Illinois (US)
1948
Harnstoffharz
        Urea resin
14,7 x 28,1 x 16,5 cm
Inv.Nr. K 1614 W

1400
Stromberg-Carlson Co.
Rochester, New York (US)
1948
Kunststoff
        Plastic
19,5 x 31,5 x 18,5 cm
Inv.Nr. K 1615 W

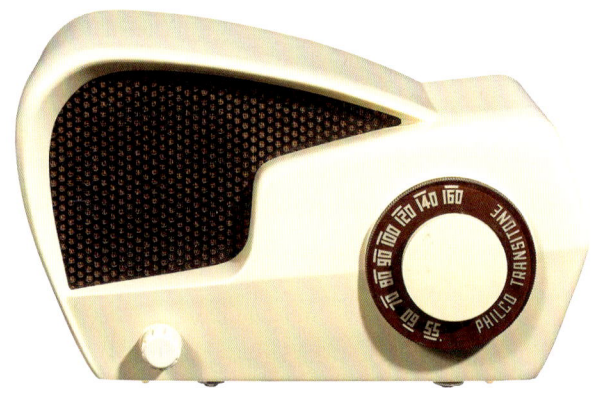

Transitone 49501 Boomerang
Philco Corp.
Philadelphia, Pennsylvania (US)
1948-49
Press-Phenolharz (Bakelit), lackiert
    Pressed phenolic resin (Bakelit), lacquered
17,5 x 29 x 15 cm
Inv.Nr. K 1350 W

Transitone 49503 Flying Wedge
Philco
Philadelphia, Pennsylvania (US)
1948-49
Kunststoff
    Plastic
17 x 30 x 19,5 cm
Inv.Nr. K 1349 W

8 K 591
RCA-Victor
New York City, New York (US)
1949
Kunststoff
        Plastic
14,8 x 21,5 x 13 cm
Inv.Nr. K 1360 W

602 A
Raymond Loewy
Emerson Radio and Phonograph Corp.
New York City, New York (US)
1949
Kunststoff
        Plastic
15,8 x 23,5 x 17,5 cm
Inv.Nr. K 1299 W

5 A 410 A Cooler-Radio
Point of Purchase Displays Inc.
Chicago, Illinois (US)
1949
Kunststoff
        Plastic
25 x 20,5 x 21 cm
Inv.Nr. K 1272 W

Super 560 W
Max Braun oHG
Frankfurt a. M. (DE)
1949-50
Press-Phenolharz (Bakelit)
        Pressed phenolic resin (Bakelit)
31 x 42 x 21 cm
Inv.Nr. K 1746

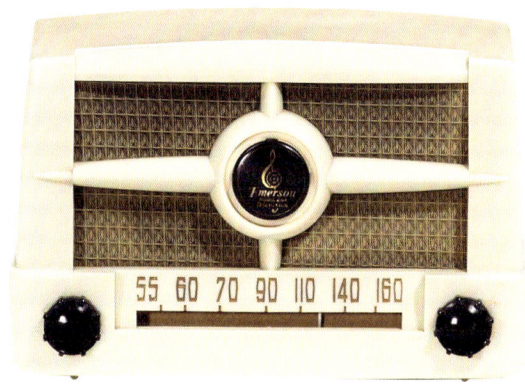

1500 H The Dynatomic
Stromberg-Carlson
Rochester, New York (US)
1949-50
Press-Phenolharz (Bakelit)
        Pressed phenolic resin (Bakelit)
19,5 x 31 x 18 cm
Inv.Nr. K 1381 W (braun / brown)
Inv.Nr. K 1382 W (rot / red)

587 A
Emerson Radio and Phonograph Corp.
New York City, New York (US)
1949-50
Kunststoff
        Plastic
14,5 x 22 x 14 cm
Inv.Nr. K 1294 W

10-135; -137 Studebaker
Carl Reynolds Jr.
Crosley & Division AVCO Manufacturing Corp.
Cincinnati, Ohio (US)
1949-51
Kunststoff, lackiert
      Plastic, lacquered
17,5 x 33 x 18 cm
Inv.Nr. K 1280 W (weiß / white, -135)
Inv.Nr. K 1281 W (grün / green, -137)

5 X 11
Motorola Inc.
Chicago, Illinois (US)
1949-51
Press-Phenolharz (Bakelit); Metall
      Pressed phenolic resin (Bakelit); metal
18 x 28 x 16 cm
Inv.Nr. K 1330 W

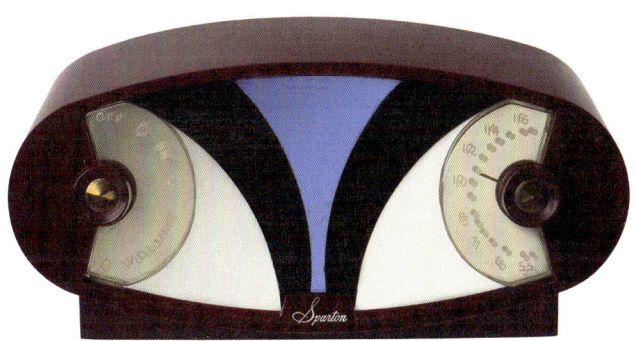

5 H 11
Motorola Inc.
Chicago, Illinois (US)
1949-52
Press-Phenolharz (Bakelit)
    Pressed phenolic resin (Bakelit)
17 x 27 x 15 cm
Inv.Nr. K 1334 W

Sparton 132 Football
Sparks-Withington Company
Jackson, Michigan (US)
1950
Kunststoff; Glas, verspiegelt
    Plastic; glass, metalized
14,7 x 32,8 x 17,5 cm
Inv.Nr. K 1618 W

70

5 X 11 U
Motorola Inc.
Chicago, Illinois (US)
1950-51
Press-Phenolharz (Bakelit)
        Pressed phenolic resin (Bakelit)
17,6 x 27,5 x 17,5 cm
Inv.Nr. K 1340 W

Super 860 W
Max Braun oHG
Frankfurt a.M. (DE)
1950-51
Holz; Press-Phenolharz (Bakelit)
        Wood; pressed phenolic resin (Bakelit)
34 x 62 x 29 cm
Inv.Nr. K 1747

Crosley 11-107 U Decorator
Carl Reynolds Jr.
Division AVCO Manufacturing Corp.
Cincinnati, Ohio (US)
1950-52
Kunststoff
        Plastic
25,5 x 33 x 19 cm
Inv.Nr. K 1289 W

T 402-F; -V
Zenith Radio Corp.
Chicago, Illinois (US)
1950-55
Kunststoff
        Plastic
15,7 x 20 x 11,5 cm
Inv.Nr. K 1399 W (grün / green, -F)
Inv.Nr. K 1400 W (rosé / pink, -V)

Manche Legenden leben länger. Auf kaum ein Produkt trifft dies mehr zu als auf die legendäre 55er Serie von Mikrofonen der US-amerikanischen Firma Shure. Die ersten kamen 1939 auf den Markt und sie werden mit aktualisierter Technik noch immer produziert – unglaubliche 77 Jahre lang. Dieses Mikrofon ist so berühmt wie seine Nutzer: Marlene Dietrich, Ella Fitzgerald, Indira Gandhi, John F. Kennedy, Martin Luther King, Frank Sinatra, Harry Truman und – allen voran – Elvis Presley sind die Prominenten, deren Bilder um die ganze Welt gingen und mit ihnen das Shure „55". Der Hersteller selbst wirbt daher selbstbewusst und sinnträchtig mit dem Slogan: „The Microphone that needs no name", also „Das Mikrofon, das keinen Namen braucht". Ein gewisser Benjamin Baumzweiger, der seinen Namen später in Ben Bauer abänderte, entwickelte das erste 55er in den frühen 1937er Jahren. Man sieht dem Design auch heute noch deutlich seine zeitlichen Wurzeln an: Die Streamline-Moderne und das Machine Age haben ihre Spuren hinterlassen, aber selbst damals hätte es auch gut in der Science Fiction auftauchen können. Es ist so futuristisch gestaltet, dass es seine Modernität nie verloren hat.

Shure Microphone PE 55 Unidyne II;
55 S Unidyne, „Baby Unidyne"
Ben Bauer
Shure Brothers Company
Evanston, Illinois (US)
1951; 1951-57
Aluminium; Metall; Kunststoff
        Aluminium; metal; plastic
19,5 x 5,8 x 8 cm
Inv.Nr. K 1733 W
Inv.Nr. K 1734 W

Some legends live longer than others: there's hardly another product to which this saying would apply better than to the legendary 55 range of microphones from US company Shure. The first models were launched in 1939 and, fitted with current technology, the design is still being produced today – making for an incredible 77 years of production! This microphone is as famous as its users: Marlene Dietrich, Ella Fitzgerald, Indira Gandhi, John F. Kennedy, Martin Luther King, Frank Sinatra, Harry Truman and, above all, Elvis Presley are among the stars whose pictures have gone around the world, and with those pictures, the Shure "55". Evoking these images, the manufacturer uses the confident advertising slogan: "The microphone that needs no name." Benjamin Baumzweiger, who later changed his name to Ben Bauer, developed the first "55" in 1937. The current design still clearly harks back to its origins: streamline modernism and the machine age have left their traces while, back in the late 1930s, the design could also have featured in science fiction. The microphone's design is so futuristic that it has never lost its modern feel.   RB

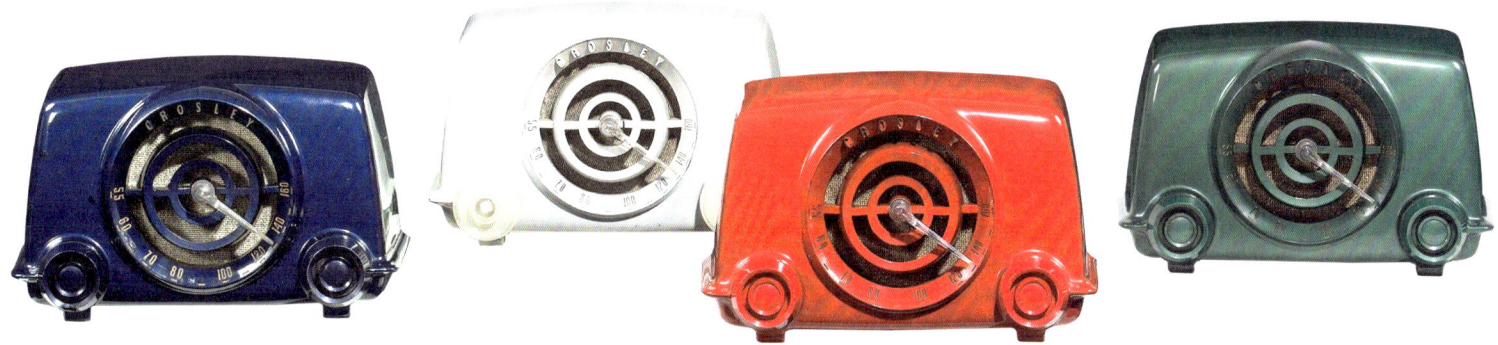

Crosley 11-100 U bis -105 U Center Bullseye
Division AVCO Manufacturing Corp.
Cincinnati, Ohio (US)
1951
Press-Phenolharz (Bakelit), lackiert
        Pressed phenolic resin (Bakelit), lacquered
14,5 x 26 x 16 cm
Inv.Nr. K 1284 W a + b (blau / blue, -101 U)
Inv.Nr. K 1285 W a + b (weiß / white, -100 U)
Inv.Nr. K 1286 W a + b (rot / red, -103 U)
Inv.Nr. K 1288 W (dunkelgrün / dark green, -102 U)
Inv.Nr. K 1287 W (dunkelgrün / dark green, -102 U)
Inv.Nr. K 1604 W (grün, ohne Abb. / green, without fig., D 10 GN)
Inv.Nr. K 1605 W (hellgrün, ohne Abb. / light green, without fig., -105 U)
Inv.Nr. K 1611 W (schwarz, ohne Abb. / black, without fig., -104 U)

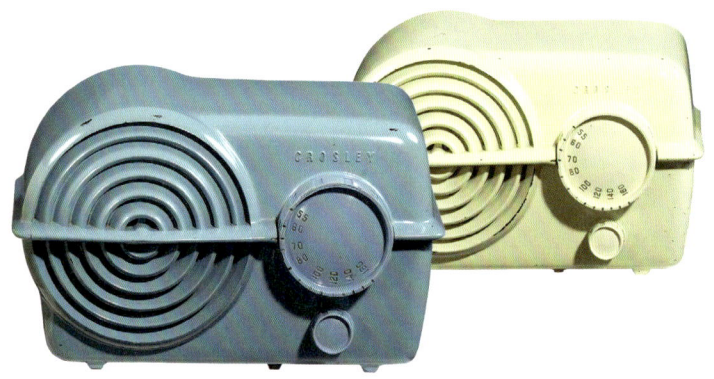

Crosley 11-114 U; -119 U Serenader, Left Bullseye   52R
Division AVCO Manufacturing Corp.   Motorola Inc.
Cincinnati, Ohio (US)   Chicago, Illinois (US)
1951   1951-52
Press-Phenolharz (Bakelit), lackiert   Kunststoff
        Pressed phenolic resin (Bakelit), lacquered   Plastic
14,5 x 23,5 x 14,5 cm   15,2 x 23 x 12,5 cm
Inv.Nr. K 1282 W (weiß / white, -114 U)   Inv.Nr. K 1328 W
Inv.Nr. K 1283 W (blau / blue, -119 U)

Crosley D 25-BE; -CE
Division AVCO Manufacturing Corp.
Cincinnati, Ohio (US)
1951-53
Press-Phenolharz (Bakelit), lackiert
        Pressed phenolic resin (Bakelit), lacquered
19,5 x 34 x 18 cm
Inv.Nr. K 1276 W (grün / green, -CE)
Inv.Nr. K 1277 W (grau / grey, -BE)

K 412 R Crest, „Owl Eye"
Zenith Radio Corp.
Chicago, Illinois (US)
1952-53
Press-Phenolharz (Bakelit)
        Pressed phenolic resin (Bakelit), lacquered
11,5 x 17 x 10,5 cm
Inv.Nr. K 1395 W

Crosley E 15-BE; -CE
Division AVCO Manufacturing Corp.
Cincinnati, Ohio (US)
1953
Kunststoff, lackiert
        Plastic, lacquered
18 x 33 x 20 cm
Inv.Nr. K 1274 W (blau / blue, -BE)
Inv.Nr. K 1275 W (grün / green, -CE)

21 RD 253 A Automatic,
Leonardo Spezial Kombinationstruhe
N. V. Philips' Gloeilampenfabrieken
Eindhoven (NL)
1953-57
Tischlerplatte; Birke, furniert; Nussbaum, furniert;
Kunststoff; Messing; Textil
        Wood-core plywood; birch veneer; walnut
        veneer; plastic; brass; textile
97/125 x 78 x 50/65 cm
Inv.Nr. A 1725

309 U Talisman
Tesla
Prag (ČSR)
1953-58
Press-Phenolharz (Bakelit)
        Pressed phenolic resin (Bakelit)
18 x 31,5 x 16,5 cm
Inv.Nr. K 1383 W

M 510 Y
Zenith Radio Corp.
Chicago, Illinois (US)
1954-55
Kunststoff
        Plastic
15 x 29,5 x 15 cm
Inv.Nr. K 1393 W

Zwezda 54
ZH-68
Moskau (UdSSR)
1954-56
Kunststoff
      Plastic
57 x 38 x 26 cm
Inv.Nr. K 1312 W

123
Tremont
(US)
1954-56
Kunststoff
      Plastic
15,3 x 26 x 14,5 cm
Inv.Nr. K 1386 W

R 521 Y
Zenith Radio Corp.
Chicago, Illinois (US)
1954-57
Kunststoff
        Plastic
14,2 x 33 x 18 cm
Inv.Nr. K 1396 W

Tischsuper / Table super G 11
Hans Gugelot
Max Braun oHG
Frankfurt a.M. (DE)
1955
Ahorn; Kunststoff
        Maple; plastic
39 x 34 x 53,9 cm
Inv.Nr. K 668

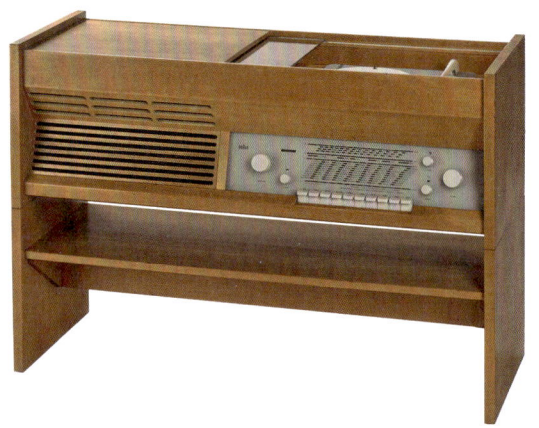

Tischsuper / Table super TSG
Hans Gugelot, Helmut Müller-Kühn
Max Braun oHG
Frankfurt a.M. (DE)
1955
Birke; Glas; Kunststoff
    Birch; glass; plastic
36,5 x 32 x 58 cm
Inv.Nr. K 651

PKG „Langer Heinrich"
Hans Gugelot
Max Braun oHG
Frankfurt a.M. (DE)
1955
Tischlerplatte; Birke, furniert; Kunststoff
    Wood-core plywood; birch veneer; plastic
66,5 x 94,5 x 40 cm
Inv.Nr. A 1648

56 R 4
Motorola Inc.
Chicago, Illinois (US)
1955-56
Kunststoff
       Plastic
15 x 25 x 15 cm
Inv.Nr. K 1325 W

56 B
Motorola Inc.
Chicago, Illinois (US)
1955-56
Kunststoff
       Plastic
24 x 26,5 x 9 cm
Inv.Nr. K 1326 W

5 C 11 E
Motorola Inc.
Chicago, Illinois (US)
1955-59
Kunststoff
    Plastic
13,6 x 31 x 15 cm
Inv.Nr. K 1329 W

85515
Motorola Inc.
Chicago, Illinois (US)
1955-59
Kunststoff
    Plastic
24 x 26,5 x 9 cm
Inv.Nr. K 1341 W

5 P 31 A
Motorola Inc.
Chicago, Illinois (US)
1956-57
Kunststoff
        Plastic
22 x 26,5 x 8 cm
Inv.Nr. K 1333 W

558 P 4
Westinghouse Electric. Corp.
Sunbury, Pennsylvania (US)
1956-57
Kunststoff; Metall
        Plastic; metal
17 x 22 x 8,3 cm
Inv.Nr. K 1388 W

868
Emerson Radio and Phonograph Corp.
New York City, New York (US)
1956-57
Kunststoff
        Plastic
25 x 25,5 x 10
Inv.Nr. K 1293 W

A 510 Y
Zenith Radio Corp.
Chicago, Illinois (US)
1956-60
Kunststoff
        Plastic
14,8 x 30,5 x 15,5 cm
Inv.Nr. K 1401 W

Atelier 1, Mono Steuergerät mit Lautsprecher /
Controll unit with speaker L 1
Dieter Rams
Max Braun oHG
Frankfurt a.M. (DE)
1957
Tischlerplatte; Rüster, teilweise lackiert
    Wood-core plywood; elm, partly lacquered
29,8 x 58 x 29 cm
Lautsprecher / speaker 23,8 x 58 x 29 cm
Inv.Nr. K 1731 a + b

Phonosuper SK 5 „Schneewittchensarg"
Hans Gugelot, Dieter Rams
Max Braun oHG
Frankfurt a.M. (DE)
1958
Holz; Rüster; Metall; Polymethylmethacrylat
(Acrylglas)
    Wood; elm; metal; polymethyl methacrylate
    (acrylic glass)
24 x 58 x 29 cm
Inv. Nr. A 1600

Transistor 2
Dieter Rams
Max Braun oHG
Frankfurt a.M. (DE)
1958
Polystyrol; Metall; Leder
        Polystyrene; metal; leather
21 x 30 x 9,5 cm
Inv.Nr. K 1264 W

SABA Meersburg Automatic 9
Schwarzwälder-Apparate-Bau-Anstalt
August Schwer Söhne GmbH
Villingen (DE)
1958-59
Nussbaum; Kunststoff
        Walnut; plastic
38 x 63 x 30 cm
Inv.Nr. K 1738

Kleinsuper / Small super SK 2-2
Artur Braun, Fritz Eichler
Max Braun oHG
Frankfurt a.M. (DE)
1959
Kunststoff; Holz
        Plastic; wood
15,5 x 23,7 x 13 cm
Inv.Nr. K 1265 W
Inv.Nr. K 625 b (ohne Abb. / without fig.)

HM 6
Herbert Hirche
Max Braun oHG
Frankfurt a.M. (DE)
1959
Spanplatte; Nussbaum, furniert; Glas; Kunststoff
        Particle board; walnut veneer; glass; plastic
72 x 133,5 x 38,5 cm
Inv.Nr. K 1720

T 202
Trav-Ler Manufacturing Corp.
Chicago, Illinois (US)
1959
Kunststoff
        Plastic
15,5 x 24,5 x 15,5 cm
Inv.Nr. K 1385 W

5 T 11 R
Motorola Inc.
Chicago, Illinois (US)
1959
Kunststoff
        Plastic
16,5 x 30 x 14 cm
Inv.Nr. K 1335 W

A 1 R 46
Motorola Inc
Chicago, Illinois (US)
1959-60
Kunststoff
        Plastic
16,2 x 23 x 12,5 cm
Inv.Nr. K 1342 W

C 1 N 3 Telechron
Motorola Inc.
Chicago, Illinois (US)
1959-61
Kunststoff
        Plastic
14,5 x 28,5 x 16,5 cm
Inv.Nr. K 1336 W

C 2 IBKJ
Motorola Inc.
Chicago, Illinois (US)
1959-61
Kunststoff
     Plastic
15 x 28,5 x 14 cm
Inv.Nr. K 1339 W

C 8 W 3
Motorola Inc.
Chicago, Illinois (US)
1959-61
Kunststoff
     Plastic
14,4 x 28,5 x 17 cm
Inv.Nr. K 1337 W

RCS 9 Steuergerät / Controll unit
Dieter Rams
Braun AG
Kronberg (DE)
1961
Holz; Kunststoff; Aluminium
        Wood; plastic; aluminium
21 x 56,6 x 28,5 cm
Inv.Nr. K 925

Kleinsuper / Small super SK 25
Artur Braun, Fritz Eichler
Max Braun oHG
Frankfurt a.M. (DE)
1961
Kunststoff; Holz
        Plastic; wood
14,5 x 13 x 23,5 cm
Inv.Nr. K 625 a

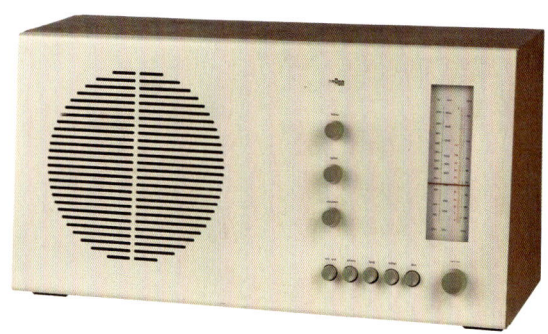

Tischsuper / Table super RT 20
Dieter Rams
Max Braun oHG
Frankfurt a.M. (DE)
1961
Buche; Kunststoff
      Beech; plastic
25,8 x 50 x 19 cm
Inv.Nr. K 975

Audio 1, TC 40 mit Lautsprecher / with speaker L 50
Dieter Rams
Braun AG; Max Braun oHG
Frankfurt a.M. (DE)
1962; 1961
Kunststoff; Metall; Aluminium; Gummi
Holz; Stahlstäbe; Metallgitter
      Plastic; metal; aluminium; rubber
      Wood; steel bars; metal grille
16 x 65 x 28 cm
61 x 65,5 x 28,5 cm
Inv.Nr. K 1620 W
Inv.Nr. K 1621 W (Lautsprecher / speaker)

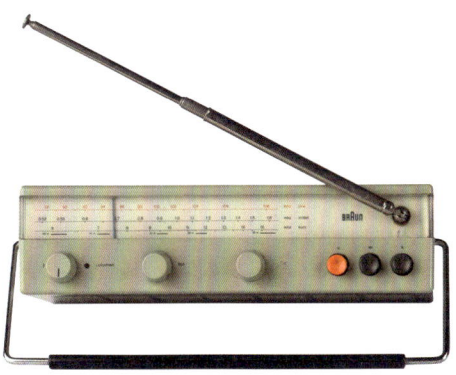

T 521
Dieter Rams
Braun AG
Kronberg (DE)
1962
Kunststoff; Metall; Polymethylmethacrylat
(Acrylglas)
        Plastic; metal; polymethyl methacrylate
        (acrylic glass)
17 x 23 x 6 cm
Inv.Nr. K 688

T 520
Dieter Rams
Braun AG
Frankfurt a.M. (DE)
1962
Kunststoff; Stahl
        Plastic; steel
19,5 x 25 x 6 cm
Inv.Nr. K 1267 W

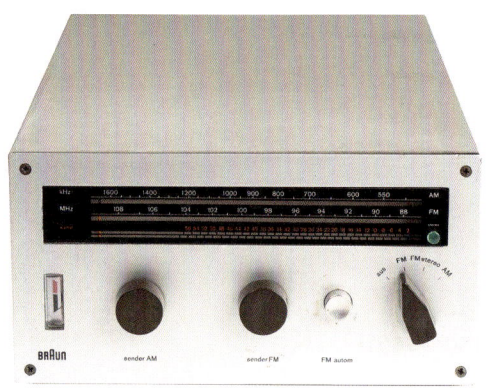

K 615 B
Zenith Radio Corp.
Chicago, Illinois (US)
1963
Kunststoff
      Plastic
22,5 x 27,5 x 16,5 cm
Inv.Nr. K 1394 W

CE 16 Radioempfänger / Radio receiver
Dieter Rams
Braun AG
Frankfurt a.M. (DE)
1964-65
Stahlblech; Aluminium
      Steel panel; aluminium
11 x 20,5 x 34 cm
Inv.Nr. K 1742

Das Klappradio „Cubo" des italienischen Radio- und Fernsehherstellers Brionvega weist einen skulpturalen Charakter auf. Im geschlossenen Zustand verbirgt der Quader aus glänzendem ABS-Kunststoff seine Funktion. Selbst der verchromte Tragegriff und die Antenne können eingefahren werden, sodass sie sich der stereometrischen Form unterordnen. Erst nach dem Aufklappen teilt sich der Quader in zwei Kuben und offenbart in seinem Inneren Lautsprecher, Skala und Bedienelemente. „Cubo" war in den Grundfarben Rot, Gelb und Blau sowie in Weiß und Schwarz erhältlich. Brionvega setzte seit den 1960er Jahren einen Schwerpunkt auf das Produktdesign, um seine technisch ausgefeilten Geräte von der Massenware abzugrenzen. Dieses Ziel erreicht das Modell „TS 502" durch seine Nominierung 1964 für einen Compasso d´Oro und 1965 durch die Goldmedaille der Biennale für Industriedesign in Ljubljana. Richard Sapper und Marco Zanuso kreierten mit dem „TS 502" einen zeitlosen Meilenstein des italienischen Designs, der in Reeditionen über Jahrzehnte neuaufgelegt und an den technischen Fortschritt angepasst wurde. Die dritte Edition von 1977, das „TS 505", mit einer verbesserten technischen Ausstattung und einer Neugestaltung der Bedienelemente im Innern ist ebenfalls im Bestand des MAKK erhalten.

TS 502; TS 505 „Cubo"
Richard Sapper, Marco Zanuso
Brionvega S.p.A.
Mailand (IT)
1964; 1977
ABS-Kunststoff; Metall
        ABS-plastic; metal
13 x 23 x 13 cm
Inv.Nr. K 662 (rot, ohne Abb. / red, without fig.)
Inv.Nr. K 1268W (gelb / yellow)
Inv.Nr. K 1269 W (rot / red)
Inv.Nr. K 1270 W (schwarz / black)
Inv.Nr. K 1271 W (weiß / white, TS 505)

The foldable "Cubo" from Italian radio and television set manufacturer Brionvega has a sculptural character. When folded, the cube, with its shiny casing made from ABS plastic, does not give away its function. Even the chromed handle and the aerial can be retracted so as not to disturb the cubic volume. Only after unfolding, "Cubo" divides into two cubes, revealing speaker, dial and controls. "Cubo" was available in the base colours red, yellow and blue, and in white and black. Since the 1960s, Brionvega has focused on product design to differentiate its technologically sophisticated devices from mass-produced products. This aim was achieved with the "TS 502" design's nomination for the 1964 Compasso d'Oro. In 1965, the "TS 502" was awarded the gold medal at the Ljubljana industrial design biennial. With the "TS 502", Richard Sapper and Marco Zanuso created a timeless milestone of Italian design, which, over decades, was reproduced and updated to reflect technological progress. The third re-edition from 1977, the "TS 505", with improved technological specification and redesigned control elements, is also included in the MAKK collection.   IB

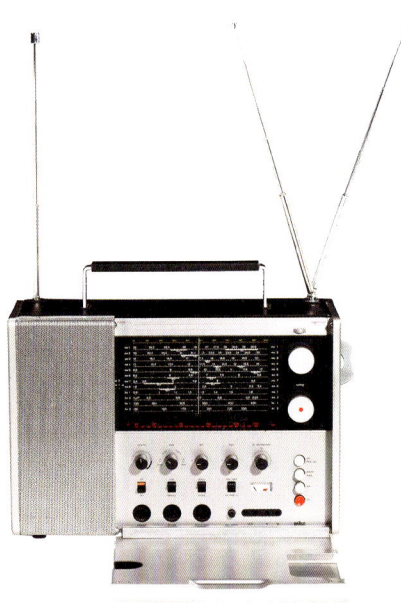

T 1000 Weltempfänger
Dieter Rams, 1963
Braun AG
Frankfurt a.M. (DE)
1964
Metall; Kunststoff
    Metal; plastic
30 x 36 x 36 cm
Inv.Nr. K 1266 W

Dieter Rams ließ sich bei seinem Entwurf für den Braun „Weltempfänger T 1000" von der Vernunft leiten, die dem Entwurfskonzept der HfG Ulm für Braun-Radios zugrunde lag. Er teilt die Flächen in eine aufrechte Lautsprecherzone mit Rasterlochung, eine horizontale, stark differenzierte Skala und eine Zone mit klar angeordneten Bedienelementen. Durch den Einsatz von Farbe hebt er auf der Skala und an den Bedienelementen den UKW-Empfang hervor. Diese konsequente Gestaltung erlaubt einen optimalen Gebrauch des hochkomplexen Mehrfrequenzradios. Das Gerät war in der Lage praktisch alle weltweit gesendeten Frequenzen zu empfangen und diente mit Zubehörteilen auch als Navigationsinstrument. Eine Schutzklappe vor Skala und Bedienelementen erlaubte einen sicheren Transport und enthält ein Fach für die umfangreiche Bedienungsanleitung. Das kühle silberfarbene Metall in Kombination mit Weiß, Schwarz und einem sparsamen Farbakzent lassen das Gerät bis heute modern erscheinen. Das Modell von 1963 war mit einer weißen Skala versehen. Der „T 1000" ist das erste Radiogerät mit der Bezeichnung „Weltempfänger" und prägte alle nachfolgenden Apparate dieser Gattung in Bezug auf Design und technische Ausstattung. Zugleich markiert es auch einen Endpunkt, da es sich um das letzte tragbare Radiomodell aus dem Hause Braun handelt.

In his design for the Braun "T 1000 Weltempfänger" (World Receiver), Dieter Rams followed the rationale on which HfG Ulm had based its design concept for Braun radios. He divided the surfaces into a vertical perforated speaker zone, a very detailed horizontal tuning dial and a clearly laid-out control panel. Using colour, he emphasised FM reception in both the tuning dial and the control elements. This consistent design ensured great usability for this highly complex, multi-frequency radio. The device was able to receive practically all frequencies used around the world. With additional accessories, the radio could also be used as a navigation device. A protective flap to cover the dial and control elements allowed safe transport and also contained a storage compartment for the manual. Thanks to the cool, silver-effect metal combined with black and white and the considered use of colour accents, the device still has a modern look and feel. The 1963 model featured a white tuning dial. The "T 1000" was the first radio to bear the title 'Weltempfänger', and shaped all successive devices of this kind in terms of design and technical configuration. At the same time, it represents the end of an era, as it was the last portable radio from Braun.   IB

Audio 2 TC 45-1 mit Lautsprechern / with speakers
L 450
Dieter Rams
Braun AG
Frankfurt a.M. (DE)
1965
Metall; Kunststoff; Polymethylmethacrylat
(Acrylglas)
Spanplatte, beschichtet; Aluminium
        Metal; plastic; polymethyl methacrylate
        (acrylic glass)
        Particle board, coated; aluminium
16,8 x 65 x 28,5 cm
Lautsprecher je / each speaker 28 x 47 x 10,5 cm
Inv Nr. K 767

RR126 OFST Deposito
Achille Castiglioni, Pier Giacomo Castiglioni
Brionvega S.p.A.
Mailand (IT)
1965-66
Holz, lackiert; Aluminium
        Wood, lacquered; aluminium
73 x 122 x 37,5 cm
Inv.Nr. K 1239 W (braun / brown)
Inv.Nr. K 1240 W (rot / red)

CE 500 Radioempfänger / Radio receiver
Dieter Rams
Braun AG
Kronberg (DE)
1967
Stahl; Aluminium; Kunststoff
        Steel; aluminium; plastic
11 x 26,5 x 33,5 cm
Inv.Nr. K 1019 b

Metropolitan
Blaupunkt-Werke GmbH
Hildesheim (DE)
1967-68
Tischlerplatte; Kanadisch-Nussbaum, furniert;
Stahlrohr; Polyvinylchlorid (Renolit)
        Wood-core plywood; canadian walnut ve-
        neer; tubular steel; polyvinyl chlorid (Renolit)
73 x 157 x 50 cm
Inv.Nr. A 2062

TR 1829
Sony Corp.
Tokio (JP)
1967-70
Kunststoff; Metall
        Plastic; metal
12,5 x DM 7,5 cm
Inv.Nr. K 1373 W

Braun Buchlabor im Schuber / Book laboratory with
slipcase
Heinz Saucke, Jules Stauber
Druck / Print: Johannes Weisbecker
Frankfurt a. M. (DE)
1969
Papier; Elektronik
        Paper; electronics assembly
32,5 x 35 x 4,5 cm
Inv.Nr. K 1740 W

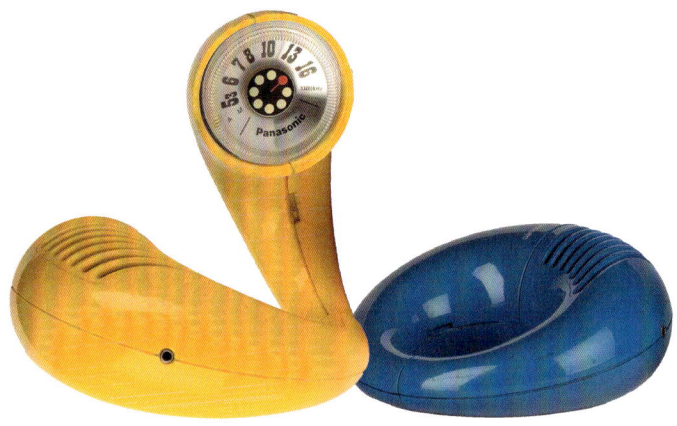

Panasonic R 72 Toot-a-loop
Matsushita Electric Industrial Co. Ltd.
Kadoma (JP)
1969-72
ABS-Kunststoff; Metall
        ABS-plastic; metal
7/13 x DM 15,5 cm
Inv.Nr. K 1345 W a (gelb / yellow)
Inv.Nr. K 1345 W b (blau / blue)

„It's an S it's an O, it's a crazy radio!", mit diesem Slogan bewarb Matsushita Electric Industrial sein tragbares Radio auf dem internationalen Markt. Die Werbung bezieht sich auf seine im Radiodesign völlig neue Gestalt, die geschlossen wie ein O erscheint und geöffnet zu einer S-Form wird. Werbeplakate aus den frühen 1970ern zeigen ein junges Mädchen mit demonstrativ zum Betrachter ausgestrecktem Handgelenk, an dem das Radio wie ein Armreif hängt. Jedoch ist das Gerät zu schwer um tatsächlich über längere Zeit bequem auf diese Weise mitgeführt zu werden. Tatsächlich ist es vorstellbar, dass es am Lenker eines Fahrrads oder an einer Handtasche befestigt wurde. Da sich die Lautsprecheröffnungen und der Lautstärkenregler an der Außenseite befinden, ist es im geschlossenen Zustand möglich Rundfunk zu hören. Lediglich zum Verstellen des Senders ist es nötig die beiden Hälften des Geräts zu verdrehen. An einer der kreisrunden Innenseiten befindet sich die Skala, die zugleich zur Amplitudenmodulation und damit zur Sendereinstellung dient. Der japanische Hersteller erreichte mit seinem kostengünstigen und farbintensiven „R72 Toot-a-loop" besonders die junge Generation. Dazu trugen auch die mitgelieferten Abziehbilder bei, die der Personalisierung des Spaßgerätes dienten.

„Wear it. Swing It. Twist it. You can even listen to it!"

"It's an S it's an O, it's a crazy radio!" – with this slogan, Matsushita Electric Industrial advertised its portable radio on the international market. The campaign referred to the radio's radically new shape, which looked like an O when closed and turned into an S when opened. Advertising posters from the early 1970s show a young girl stretching her arm towards the viewer, wearing the radio on her wrist like a bracelet. However, the device is too heavy to be worn comfortably like this over a prolonged period of time. But it could easily have been attached to a bicycle's handlebars or to a handbag. As the speaker openings and the volume control are positioned on the exterior, it is possible to listen to the radio when it's closed. The two halves of the device needed to be twisted apart only when wanting to change stations. The dial display, which is also used for amplitude modulation and hence for tuning into another station, is embedded into one of the circular interior surfaces. The reasonably priced and colourful "R72 Toot-a-loop" was particularly popular among young people, a fact that was also emphasised by the stickers that came with the radio, which could be used to personalise this fun product.

"Wear it. Swing it. Twist it. You can even listen to it!"   IB

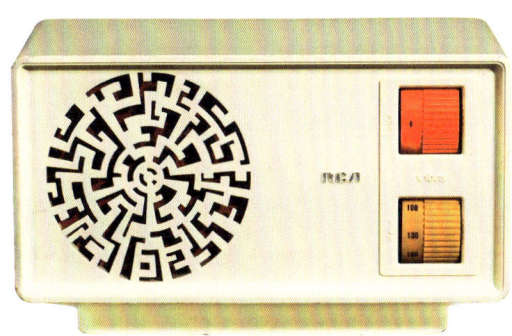

RVA 701 Y Mandala
RCA-Victor
New York City, New York (US)
1969-71
Kunststoff
        Plastic
13,4 x 23,5 x 13 cm
Inv.Nr. K 1357 W

3203 FET Musikstudio / Music studio
Wega Radio GmbH
Fellbach (DE)
1970
Schleiflack; Kunststoff
        Varnish; plastic
18,5 x 71,5 x 33 cm
Inv.Nr. K 778

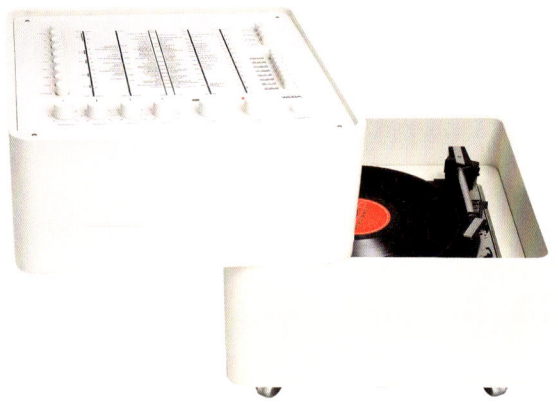

Wega Stereobar 3300
Verner Panton
Wega Radio GmbH
Fellbach (DE)
1970
Kunststoff; Metall; Holz; Schleiflack
          Plastic; metal; wood; varnish
44 x 42,5 x 42,5 cm
Inv.Nr. K 1243 W

„Die Technik hält, was die Form verspricht", wirbt die Wega-Radio GmbH aus Fellbach bei Stuttgart in den 1960er Jahren. Aus der Wohlgestalt des Apparats soll der Kunde auf die nicht sichtbare Leistungsfähigkeit der technischen Komponenten schließen. Der Slogan findet sich in Reklameanzeigen der von Verner Panton entworfenen „Stereobar 3300". Das HiFi-Gerät besteht aus zwei übereinander angebrachten Kompartimenten, die geschlossen einen Würfel mit abgerundeten Kanten bilden. Die beiden Fächer sind an einer Kante miteinander verbunden, so dass sie sich gegeneinander verschieben und somit öffnen lassen. Im oberen Fach befinden sich das Radioempfangsgerät und der Verstärker mit runden Drehreglern, Knöpfen und vertikalen Skalen. Das Bedienfeld ähnelt in seiner klaren Anordnung und den roten Farbakzenten den Geräten der Braun AG. Im geschützten unteren Fach ist der Plattenspieler eingelassen. Der Würfel ruht auf vier Rollen, die es erlauben die „Stereobar" ohne Aufwand frei im Raum zu platzieren. Der dänische Designer entwarf 1965 einen Prototyp der „Stereobar" in zylindrischer Form mit drei Fächern. Dieses Prinzip der übereinander angeordneten zylindrischen Kompartimente verwendete Panton bereits 1963 für sein namensverwandtes Möbel „BarBoy". Die zylindrische Form setzte sich beim HiFi-Gerät im Gegensatz zum „BarBoy" nicht durch, da es zu Schwierigkeiten mit genormten elektronischen Einbauelementen kam.

"The technology delivers what the form promises" was the 1960s advertising slogan by Wega-Radio GmbH. The company, based in Fellbach near Stuttgart, wanted its customers to take the device's beautiful form as an indicator for the invisible high-performance technology inside. The slogan can be found in print adverts for the Stereobar 3300, which was designed by Verner Panton. The HiFi device consists of two modules fitted one above the other. When closed, the two modules form a cube with rounded edges. The modules are connected at one edge so that they can be swung out in opposite directions to open the cube. The upper module houses the radio receiver and the amplifier with rotary controls, buttons and vertical scale bar. With its clear layout and red colour accents, the control panel is reminiscent of devices from Braun AG. The turntable is embedded in the protected lower module. The cube is mounted on four castors, allowing the Stereobar to be easily placed anywhere in the room. In 1965 Danish designer Verner Panton developed a Stereobar prototype featuring a cylindrical form with three modules. Panton had already used the principle of cylindrical modules arranged one above the other in his 1963 BarBoy mobile storage furniture. The cylindrical form was eventually discarded for HiFi designs because it was difficult to fit standardised electronic components into this shape.

IB

Panasonic R 70 Panapet
Matsushita Electric Industrial Co. Ltd.
Kadoma (JP)
1970
Kunststoff; Metall
        Plastic; metal
10,5 x DM 11 cm
Inv.Nr. K 1344 W a - f

ICF 111 B Solid State Sports 11
Sony Corp.
Tokio (JP)
1970
Stahl; Aluminium; Kunststoff
        Steel; aluminium; plastic
18,5 x 22 x 5,5 cm
Inv.Nr. K 991

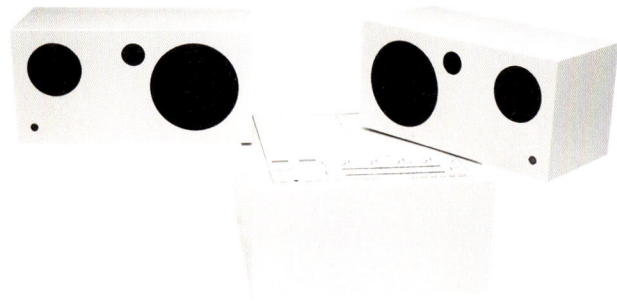

Cockpit 250 SK mit Lautsprechern / with speakers
 L 420-1
Dieter Rams
Braun AG
Kronberg (DE)
1970; 1972
Kunststoff; Polymethylmethacrylat (Acrylglas)
Kunststoff; Aluminium
        Plastic; polymethyl methacrylate
        (acrylic glass)
        Plastic; aluminium
17,5 x 57 x 35,5 cm
Lautsprecher je / each speaker 21,5 x 32 x 17 cm
Inv.Nr. K 780 a + b

RR 130 Totem
Mario Bellini
Brionvega S.p.A.
Mailand (IT)
1971
Spanplatte, lackiert; Kunststoff; Metall
        Particle board, lacquered; plastic; metal
52,5 x 51 x 52 cm
Inv.Nr. K 1241 W

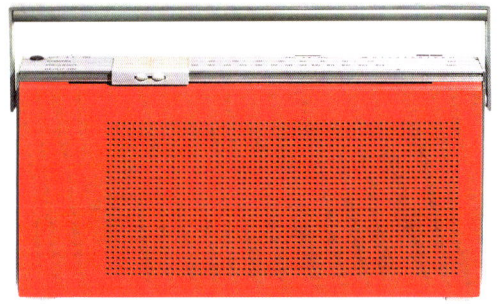

3120 HiFi
Wega Radio GmbH
Fellbach (DE)
1971-75
Metall; Kunststoff
        Metal; plastic
11,5 x 60 x 30 cm
Inv.Nr. K 1686

Beolit 400
Jacob Jensen
Bang & Olufsen
Struer (DK)
1971-75
Kunststoff; Aluminium
        Plastic; aluminium
21,5 x 36 x 7,5 cm
Inv.Nr. K 1254 W

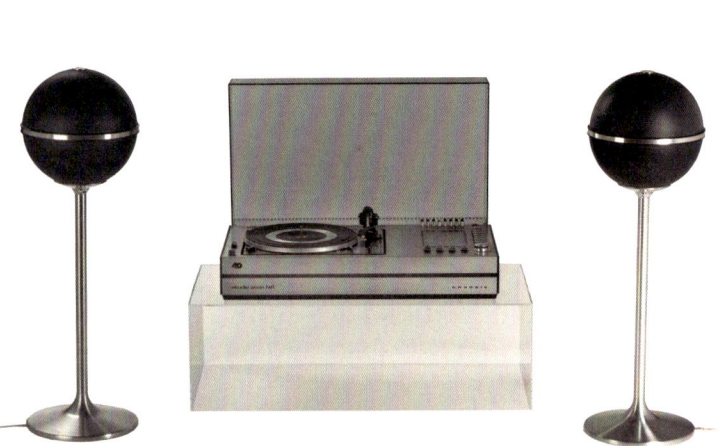

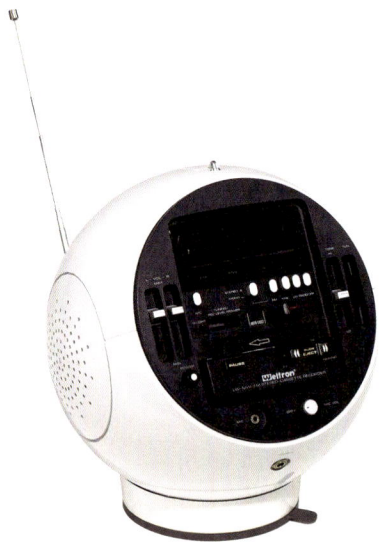

4D Stereo Studio 2000 HiFi
mit Lautsprechern / with speakers Audiorama 5000
Grundig AG
Fürth (DE)
1972; 1975
Metall, lackiert; Aluminium; Kunststoff
     Metal, lacquered; aluminium; plastic
18 x 64,5 x 39 cm
Lautsprecher je / each speaker 80,5 x DM 26 cm
Inv.Nr. K 1242 W a + b

2004 Prinz Sound Stereo
Weltron Co. Inc.
Durham, North Carolina (US)
1973
ABS-Kunststoff; Metall; Polymethylmethacrylat
(Acrylglas)
     ABS-plastic; metal; polymethyl methacrylate
     (acrylic glass)
17,5 x 27,5 x 26,5 cm
Inv.Nr. K 1387 W

137

138

Audio 308 S
Dieter Rams
Braun AG
Kronberg (DE)
1975
Kunststoff; Polymethylmethacrylat (Acrylglas);
Aluminium
        Plastic; polymethyl methacrylate (acrylic
        glass); aluminium
16,5 x 80 x 36 cm
Inv.Nr. K 976

ABR 21 Signal Radio
Dieter Rams, Dietrich Lubs
Braun AG
Kronberg (DE)
1978
Kunststoff; Metall
        Plastic; metal
11,5 x 6 x 18 cm
Inv.Nr. K 675

„Das will ich für Apple" soll Steve Jobs in einem ersten Gespräch von Hartmut Esslinger gefordert haben. Gemeint waren Esslingers Entwürfe für die Wega Radio GmbH beziehungsweise die Sony Corp., die das deutsche Unternehmen 1975 übernommen hatte. Schon während seines Designstudiums war Esslinger für Wega tätig. Im Jahr 1978 entwarf er das Kompakt-System „Concept 51 K" im Bausteinprinzip. Wie bei einem Großrechner sind die verschiedenen Komponenten in einzelnen Modulen übersichtlich aneinander gereiht. Links befindet sich unter einer Abdeckhaube der Plattenspieler, der von außen direkt gesteuert werden konnte, ohne die Haube zu öffnen. Dadurch wurde die klare Linie der Kompaktanlage während des Betriebs nicht gestört. Die rechte Hälfte des Gerätes wird von den drei Modulen Kassetten-Deck, Radioempfänger und Verstärker ausgefüllt, wobei jedes Modul in drei übereinander angeordnete Felder eingeteilt ist. Insgesamt handelt es sich um ein sehr flaches Gehäuse, das die Technik ‚hauteng' umhüllt. Das Magazin „HiFi Stereophonie" bezeichnete 1979 die Käufer des „Concept 51 K" als „anspruchsvolle und gleichzeitig designbewußte Musikfreunde". Das Museum of Modern Art in New York erwarb ein Modell in den 1970er Jahren für seine Designsammlung und auch Steve Jobs soll insbesondere vom „Concept 51 K" überzeugt gewesen sein. Schließlich erlangten Esslinger und sein Studio frogdesign zusammen mit Apple Weltruhm.

Concept 51 K
Hartmut Esslinger
Wega Radio GmbH
Fellbach (DE)
1978
Kunststoff
     Plastic
13,5 x 84 x 40 cm
Inv.Nr. K 951

"This is what I want for Apple", Steve Jobs is said to have demanded in his first conversation with Hartmut Esslinger. Jobs was referring to Esslinger's designs for Wega Radio GmbH and Sony Corp., the latter having acquired the German company in 1975. Esslinger had already worked for Wega while he was a design student. In 1978, he designed the modular "Concept 51 K" compact system. Like in a mainframe computer, the different components are arranged in clearly laid-out modules. On the left-hand side sits the turntable under its lid. It could be controlled without having to lift said lid, so the system's clear linear design would not be spoilt when using the device. The three modules of tape deck, radio receiver and amplifier are integrated at the device's right-hand side, with each module being divided into three zones, arranged one above the other. Overall, the casing is quite flat, closely hugging the technology inside. In 1979, the HiFi Stereophonie magazine referred to the buyers of the "Concept 51 K" as "discerning and also design-aware music lovers." In the 1970s, the Museum of Modern Art in New York acquired a "Concept 51 K" for its design collection. Steve Jobs, too, is said to have been especially won over by the Concept 51 K. Eventually, together with Apple, Esslinger and his frogdesign studio would become famous the world over.   IB

070 Blues
N. V. Philips' Gloeilampenfabrieken
Eindhoven (NL)
1978-79
Kunststoff; Metall
          Plastic; metal
8,5 x 14 x 4 cm
Inv.Nr. K 1681

Toshiba System 15 Aurex
Toshiba Corp.
Tokio (JP)
1978-80
Kunststoff
          Plastic
31 x 25,7 x 21 cm
Inv.Nr. K 1648 a - f

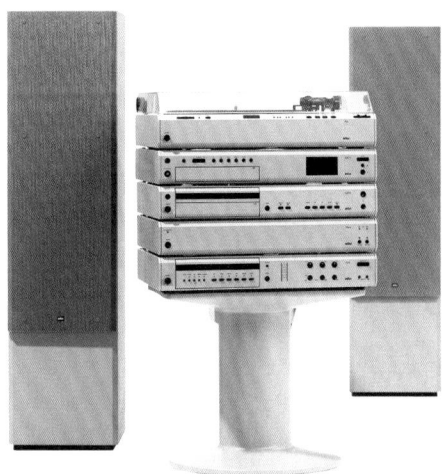

ICF 7600 A
Sony Corp.
Tokio (JP)
1982-87
Kunststoff; Metall
     Plastic; metal
11,8 x 19 x 3 cm
Inv.Nr. K 1372 W

Atelier HiFi mit Lautsprechern / with speakers
LS 150 PA und Fuß / and stand AF 1
Peter Hartwein, Dieter Rams
Braun Elektronik GmbH
Kronberg (DE)
1982-88
Metall; Kunststoff; Polymethylmethacrylat
(Acrylglas); Holz
     Metal; plastic; polymethyl methacrylate
     (acrylic glass); wood
76 x 45 x 37 cm
Lautsprecher je / each speaker 93 x 24 x 29 cm
Fuß / Stand: 94 x 24 x 29 cm
Inv.Nr. K 1238 W a + b

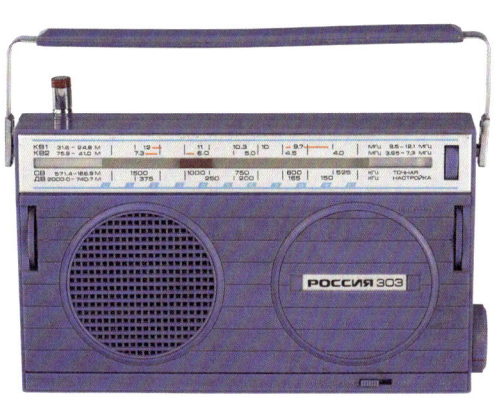

Rossiya 303
Industrial Union Poliot
Chelyabinsk (RU)
1984
Kunststoff
        Plastic
16,8 x 22,7 x 4,9 cm
Inv.Nr. K 1366 W

D 8007/32 L Roller
Graham Hinde
N. V. Philips' Gloeilampenfabrieken
Eindhoven (NL)
1986
Kunststoff
        Plastic
21 x 44 x 11,5 cm
Inv.Nr. K 1745

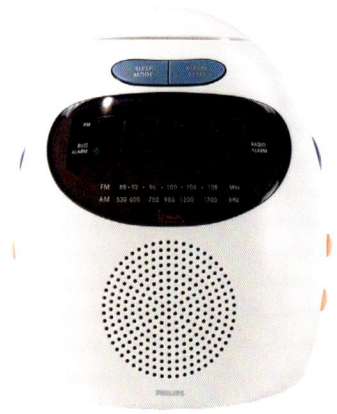

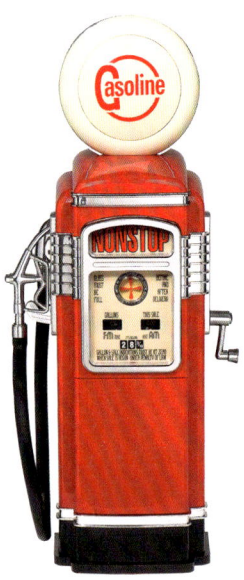

MGC 10017
Michael Graves
Philips Electronics N. V.
Eindhoven (NL)
1990-2005
Kunststoff
        Plastic
15 x 13,5 x 14 cm
Inv.Nr. K 1353 W

Gas-O-Line Petrol Pump Alarm Clock,
Collectors Edition
Marksman Polyflame Int.B.V. Product
(CN)
1991
Kunststoff; Gummi
        Plastic; rubber
58,5 x 25,5 x 14 cm
Inv.Nr. K 1346 W

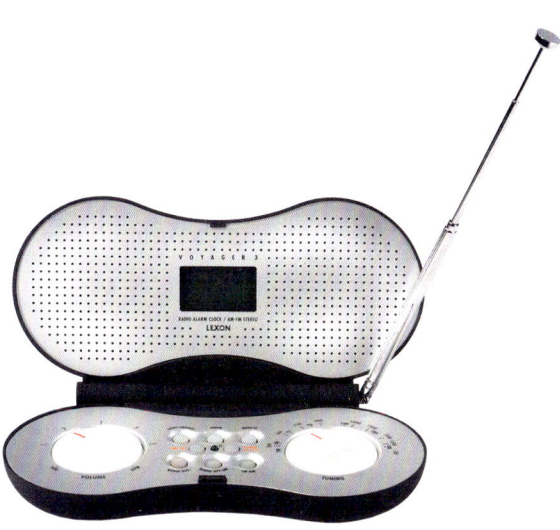

LA 39 Voyager 3
Marc Berthier
Lexon
Boulogne-Billancourt (FR)
1991- 2005
Kunststoff
        Plastic
3,8 x 16 x 8,5 cm
Inv.Nr. K 1318 W

LA 25 Times
Hervé Houplain
Lexon
Boulogne-Billancourt (FR)
1991-2005
Kunststoff; Metall
        Plastic; metal
9,8 x 9 x 9,3 cm
Inv.Nr. K 1320 W

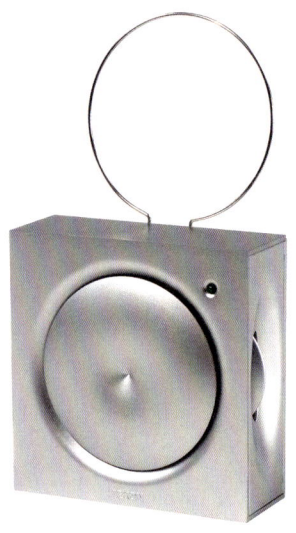

LA 50
Marc Barandard, Martin Lück, Cyril Fuchs
Lexon
Boulogne-Billancourt (FR)
1991-2005
Kunststoff
        Plastic
18,5 x 10 x 8 cm
Inv.Nr. K 1321 W

CFM 2300, My first Sony
Sony Corp.
Tokio (JP)
1994
Kunststoff; Metall
        Plastic; metal
23,3 x 35,5 x 9,5 cm
Inv.Nr. K 1374 W

20-1
Isis Electronics Ltd.
Hong Kong (CN)
1994-96
Kunststoff; Metall
        Plastic; metal
7,5 x 25,5 x 7,5 cm
Inv.Nr. K 1313 W (rot / red)
Inv.Nr. K 1314 W (verspiegelt / metalized)
Inv.Nr. K 1616 W a (rot / red, Coca-Cola)
Inv.Nr. K 1616 W b (weiß / white, Avon)

BeoSound Ouverture mit Lautsprechern / with
speakers BeoLab 2500
David Lewis
Bang & Olufsen
Struer (DK)
1994-2003
Kunststoff; Metall; Glas; Textil
        Plastic; metal; glass; textile
37 x 31,5 x 16 cm
Lautsprecher je / each speaker 37 x 26 x 16 cm
Inv.Nr. K 1237 W a - c

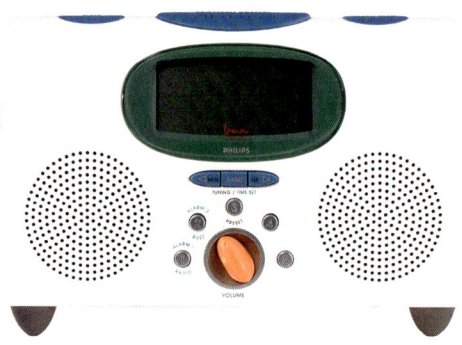

RT 201
Philippe Starck
Thomson Multimedia
Boulogne-Billancourt (FR)
1995-2000
Kunststoff; Metall; LED
          Plastic; metal; LED
19,7 x 8,5 x 4 cm
Inv.Nr. K 1384 W

MGC 20017
Michael Graves
Philips Electronics N. V.
Eindhoven (NL)
1995-2005
Kunststoff
          Plastic
13,5 x 21 x 12 cm
Inv.Nr. K 1355 W

Poe
Philippe Starck, 1994
Thomson Multimedia für / for Alessi
Boulogne-Billancourt (FR)
1996
ABS-Kunststoff
ABS-plastic
25,4 x 16,5 x 17,2 cm
Inv.Nr. K 1744

Das Radiomodell „Poe" ist ein Produkt der Zusammenarbeit von Philippe Starck, Thomson und der italienischen Designschmiede Alessi. Starck hatte von 1993 bis 1996 den Posten des Kreativchef (creative director) bei der Thomson Group inne, dem zu dieser Zeit weltweit viertgrößten Unternehmen für Unterhaltungselektronik. In einem Interview für den Designreport (11/1995) erklärte Starck den deprimierenden und seelenlosen Kästen aus schwarzem Plastik den Krieg, die den Elektronikmarkt beherrschten. Er wollte Geräte entwerfen die sich zurücknehmen und sympathisch sind. Sie sollten dem Nutzer Freude bereiten. Das grau-blaue „Poe"Radio erscheint zunächst als glatte, kühle Skulptur. Die Funktion des Lautsprechers wird durch die sich konisch erweiternde Zylinderform mit oben abgeschrägtem Abschluss betont. Die Bedienelemente und eine digitale Anzeige verbergen sich unter der schmalen Standfläche. Ein kleiner Lautstärkeregler und die Aufschrift mit der Modellbezeichnung markieren die Rückseite des Radios. Der Eindruck einer gänzlich abgeschlossenen Skulptur wird durch die Radioantenne gebrochen: Das dünne lange Kabel in farbloser Kunststoffummantelung scheint nicht recht zur eleganten Form des Geräts zu passen. Dieser Bruch erzeugt einen verspielten Charakter und lässt das Radio im Sinne Starcks freundlich erscheinen.

The "Poe" radio is the result of a collaboration between Philippe Starck, the Thomson group and Italian design company Alessi. From 1993 to 1996, Starck was creative director at the Thomson group, which was the fourth largest consumer electronics manufacturer at the time. In an interview with German magazine Designreport (11/1995), Starck declared war on the depressing and soulless black plastic boxes that dominated the electronics market. He wanted to design unobtrusive and friendly-looking devices that would bring joy to their users. At first glance, the grey-blue "Poe" radio looks like a smooth, cool sculpture. The speaker function is emphasised through the conically widening cylindrical form with its slanted, truncated surface. The control elements and a digital display are hidden in the small footprint. A small volume control and a label with the model name are fixed to the device's back. The impression of a self-contained sculpture is broken up by the aerial: the long, thin cable with its clear plastic coating does not seem to be consistent with the device's elegant form. However, this contradiction introduces a playful note, thus creating the friendly feel that Starck was after.

IB

LA 42 Tykho
Marc Berthier
Lexon
Boulogne-Billancourt (FR)
1998
Synthetischer Gummi
    Synthetic rubber
14,5 x 14 x 4 cm
Inv.Nr. K 1317 W (grün / green)
Inv.Nr. K 1319 W (schwarz / black)

ABR 314 df, Millenium Edition
Dietrich Lubs
Braun Elektronik GmbH
Kronberg (DE)
1999
Kunststoff; Metall
    Plastic; metal
8 x 17 x 3,5 cm
Inv.Nr. K 1263 W

Apollo LA 11
Hervé Houplain
Lexon
Boulogne-Billancourt (FR)
1999-2003
Kunststoff; Metall
        Plastic; metal
14 x DM 10 cm
Inv.Nr. K 1316 W a - c

Time & Weather
Philippe Starck
Oregon Scientific, Inc.
Tualatin, Oregon (US)
2003
Metall; Kunststoff
        Metal; plastic
18,5 x 18,5 x 4 cm
Inv.Nr. K 1735

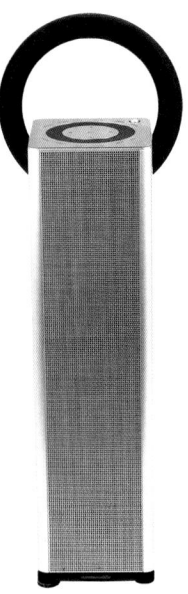

Der hohe schlanke Turm „BeoSound 3" ist das erste Radio des dänischen Herstellers Bang & Olufsen im Hochformat. Den oberen Abschluss bildet ein auffälliger schwarzer Tragegriff, der den Transport ermöglicht und an dem das Gerät mit Hilfe einer mitgelieferten Halterung an der Wand befestigt werden konnte. Die Bedienelemente mit „soft touch"-Funktion befinden sich auf der Oberseite und wiederholen in ihrer Gestaltung den schwarzen Tragering. Es handelt sich um keine Tasten, sondern Sensoren, die auf sanften Druck reagieren. Eine digitale Anzeige für Sender und Uhrzeit leuchtet im Betrieb durch das Lochraster des Lautsprechers rot auf. Im Fuß ist ein Lesegerät für SD-Karten integriert, der das Abspielen von Musikdateien ermöglicht. Die motorbetriebene Antenne fährt automatisch beim Einschalten des Radios aus. Als David Lewis den Radioturm für B & O entwarf, war er von Großbritannien wenige Jahre zuvor zum „Royal Designer for Industry" ernannt worden. Ein Titel, den seit 1936 nur 200 Designer aus dem Vereinigten Königreich zugleich beanspruchen dürfen, darunter war auch Wells Coates, der in den 1930er Jahren Radiogehäuse für Ekco entwarf. Der Brite Lewis war über 30 Jahre für B & O in Dänemark tätig und hatte sich zum Ziel gesetzt, die Technologie durch sein Design „aufzuräumen" und zu vereinfachen.

BeoSound 3
David Lewis
Bang & Olufsen
Struer (DK)
2005
Aluminium; Metall; Kunststoff
      Aluminium; metal; plastic
42,1 x 13,5 x 9 cm
Inv.Nr. K 1736

The tall and slim "BeoSound 3" tower was the first vertical radio design from Danish manufacturer Bang & Olufsen. At the top, the device features a strikingly large black handle that not only facilitates mobility but can also be used, in combination with a bracket, to mount the device to the wall. The 'soft touch' controls, whose design mirrors that of the black handle, are located on the top surface. The controls are not buttons, but sensors that respond to light pressure. The digital display shows both the selected station and the time. When the device is switched on, the red light of the display shines through the perforated speaker grid. An SD-card reader for the playback of sound files is integrated into the base. The motor-operated aerial extends automatically when the radio is switched on. A few years before David Lewis designed the radio tower for B & O, he had received the British title of Royal Designer for Industry. Introduced in the United Kingdom in 1936, only 200 designers may hold the distinction at any time. Wells Coates, who designed radio casings for Ekco in the 1930s, was also one of them. For more than 30 years, British designer Lewis worked for B & O in Denmark. With his design, he wanted to 'tidy up' and simplify the technology. IB

L Sound System
Geneva Lab
Zürich (CH)
2008
Holz; Metall; Kunststoff; Textil; Aluminium
      Wood; metal; plastic; textile; aluminium
29,1 x 44,8 x 36,5 cm, Fuß: 78,8 cm
Inv.Nr. K 1737

Cubo (Gehäuse / case)
Sonoro Audio GmbH
Köln (DE)
2008-09
Kunststoff
      Plastic
14 x 21 x 24 cm
Inv.Nr. K 1725 a
(schwarz, ohne Abb. / black, without fig.)
Inv.Nr. K 1725 b (Kirschholz, ohne Abb. /
cherrywood, without fig.)
Inv.Nr. K 1725 c (rot / red)

Konkordanzliste
Literaturverzeichnis
Abbildungsverzeichnis

Cross-Reference Index
Bibliography
Picture credits

Konkordanzliste / Cross-Reference-Index

Inv.Nr. / Kat.Nr.

Inv.Nr.	Kat.Nr.	Inv.Nr.	Kat.Nr.	Inv.Nr.	Kat.Nr.	Inv.Nr.	Kat.Nr.
A 1600	98	K 975	111	K 1250 W	44	K 1270 W	117
A 1648	88	K 976	137	K 1251 W	9	K 1271 W	117
A 1725	80	K 991	130	K 1252 W	43	K 1272 W	61
A 2062	122	K 1019 b	121	K 1253 W	46	K 1273 W	54
K 625 a	110	K 1237 W a-c	153	K 1254 W	134	K 1274 W	79
K 625 b	101	K 1238 W a-b	143	K 1257 W	39	K 1275 W	79
K 651	87	K 1239 W	120	K 1258 W	51	K 1276 W	77
K 662	117	K 1240 W	120	K 1259 W	38	K 1277 W	77
K 668	86	K 1241 W	132	K 1260 W	53	K 1278 W	49
K 675	138	K 1242 W a-b	135	K 1261 W	41	K 1279 W	49
K 688	113	K 1243 W	128	K 1263 W	158	K 1280 W	65
K 767	119	K 1244 W	27	K 1264 W	99	K 1281 W	65
K 778	127	K 1245 W	44	K 1265 W	101	K 1282 W	75
K 780 a-b	131	K 1246 W	44	K 1266 W	118	K 1283 W	75
K 793	5	K 1247 W	44	K 1267 W	114	K 1284 W a-b	74
K 925	109	K 1248 W	44	K 1268 W	117	K 1285 W a-b	74
K 951	139	K 1249 W	44	K 1269 W	117	K 1286 W a-b	74

Inv.Nr. / Kat.Nr.

Inv.Nr.	Kat.Nr.	Inv.Nr.	Kat.Nr.	Inv.Nr.	Kat.Nr.	Inv.Nr.	Kat.Nr.
K 1287 W	74	K 1306 W	36	K 1326 W	90	K 1345 W a-b	125
K 1288 W	74	K 1307 W	33	K 1327 W	40	K 1346 W	147
K 1289 W	71	K 1308 W	33	K 1328 W	76	K 1347 W	18
K 1290 W	32	K 1309 W	33	K 1329 W	91	K 1348 W	18
K 1292 W	8	K 1311 W	17	K 1330 W	66	K 1349 W	58
K 1293 W	95	K 1312 W	83	K 1331 W	24	K 1350 W	57
K 1294 W	64	K 1313 W	152	K 1332 W	24	K 1351 W	4
K 1295 W	25	K 1314 W	152	K 1333 W	93	K 1352 W	4
K 1296 W	25	K 1316 W a-c	159	K 1334 W	67	K 1353 W	146
K 1297 W	25	K 1317 W	157	K 1335 W	104	K 1355 W	155
K 1298 W	26	K 1318 W	148	K 1336 W	106	K 1356 W	3
K 1299 W	60	K 1319 W	157	K 1337 W	108	K 1357 W	126
K 1300 W	28	K 1320 W	149	K 1339 W	107	K 1360 W	59
K 1301 W	28	K 1321 W	150	K 1340 W	69	K 1361 W	35
K 1302 W	28	K 1322 W	20	K 1341 W	92	K 1362 W	35
K 1303 W a-b	47	K 1323 W	21	K 1342 W	105	K 1363 W a-b	35
K 1304 W	36	K 1324 W	6	K 1343 W	50	K 1364 W	2
K 1305 W	36	K 1325 W	89	K 1344 W a-f	129	K 1366 W	144

Inv.Nr. / Kat.Nr.

K 1367 W	37	K 1385 W	103	K 1605 W	74	K 1733 W	73
K 1368 W	29	K 1386 W	84	K 1611 W	74	K 1734 W	73
K 1369 W	19	K 1387 W	136	K 1613 W	30	K 1735	160
K 1370 W	12	K 1388 W	94	K 1614 W	55	K 1736	161
K 1371 W	22	K 1389 W a-c	34	K 1615 W	56	K 1737	162
K 1372 W	142	K 1392 W	16	K 1616 W a-b	152	K 1738	100
K 1373 W	123	K 1393 W	82	K 1617 W	53	K 1739	42
K 1374 W	151	K 1394 W	115	K 1618 W	68	K 1740 W	124
K 1375 W	10	K 1395 W	78	K 1619 W	45	K 1742	116
K 1376 W	11	K 1396 W	85	K 1620 W	112	K 1744	156
K 1377 W	23	K 1397 W	31	K 1621 W	112	K 1745	145
K 1378 W	15	K 1399 W	72	K 1648 a-f	141	K 1746	62
K 1379 W	14	K 1400 W	72	K 1681	140	K 1747	70
K 1380 W	13	K 1401 W	96	K 1686	133		
K 1381 W	63	K 1402 W	48	K 1720	102		
K 1382 W	63	K 1403 W	7	K 1725 a-c	163		
K 1383 W	81	K 1535 W	1	K 1731 a-b	97		
K 1384 W	154	K 1604 W	74	K 1732 W	52		

Literaturverzeichnis / Bibliography

Günter Friedrich ABELE: Radio-Chronik. Von der Nachkriegszeit bis zur Gegenwart. Stuttgart 2003.

Winfried ALTMAYER: Designgeschichte(n). Prof. Walter Maria Kersting, der erste Lehrer. Norderstedt 2013.

David ATTWOOD: Radio. An Appreciation. East Sussex 1997.

Martin BATTERSBY: The Decorative Thirties. London 1976.

Andreas BAUMERICH: Der Bau „sah geradezu italienisch aus." Italianismus in der rheinischen Sakralarchitektur des 20. Jahrhunderts, in: Claudia Euskirchen, Marco Kieser, Angela Pfotenhauer (Hg. / Eds.): Hörsaal, Amt und Marktplatz. Festschrift für Udo Mainzer, Sigurd Greven-Studien 6. Regensburg 2005, S. / pp. 151-160.

Catharina BERENTS: Kleine Geschichte des Designs. Von Gottfried Semper bis Philippe Starck. München / Munich 2011.

Max BILL, Schweizer Mustermesse und Schweizerischer Werkbund (Hg. / Ed.): Die gute Form. 6 Jahre Auszeichnung „Die gute Form" an der Mustermesse Basel. Winterthur 1957.

Heiner BOEHNCKE, Michael CRONE (Hg. / Eds.): Radio Radio. Studien zum Verhältnis von Literatur und Rundfunk. Frankfurt a.M. 2005.

Christian BONTEN: Kunststofftechnik für Designer. München und Wien / Munich and Vienna 2003.

Martin BÖSCH: Sony Corporation, Tokyo. ICF – 7600, 20.5.2005, http://www.shortwaveradio.ch/radio-d/sony-icf7600.htm [25.09.2015].

Kurt BRANDENBURGER: Herstellung und Verarbeitung von Kunstharzpreßmassen, 2. Auflage / 2nd edition. München und Berlin / Munich and Berlin 1938.

BRAUN AG (Hg./ Ed.): Betriebsspiegel. Braun im Urteil der Öffentlichkeit. Frankfurt a. M. 1960.

BREMER RUNDFUNKMUSEUM e.V. (Hg. / Ed.): Begleitheft zur Jubiläumsausstellung. 100 Jahre Funk seit Heinrich Hertz und 10 Jahre Bremer Rundfunkmuseum. Bremen 1992.

Gerda BREUER: Die Erfindung des modernen Klassikers. Ostfildern-Ruit 2001.

Decio Giulio Riccardo CARUGATI: Brionvega. Designing Emotion. Mailand / Milan 2003.

Germano CELANT (Hg. / Ed.): Andy Warhol. A factory, Ausst.Kat. / Exhib.cat., Kunstmuseum Wolfsburg, Kunsthalle Wien. Ostfildern 1998.

M. Michel CHEVALIER: L'Exposition Universelle de 1862. Paris 1862.

Ian CHRISTIE: Mass-Market Modernism, in: Christopher Wilk (Hg. / Ed.): Modernism. Designing a New World 1914-1939, Ausst.Kat. / Exhibit.cat., Victoria & Albert Museum. London 2006, S. / pp. 375-414.

Christine COLIN: Starck. Paris 1988.

Philip COLLINS: Radios. The Golden Age. New York 1997.

Stephan COLLOWAY: Raumdesign. Wohnen im 20. Jahrhundert. Herford 1991.

Daniel CRESPY, Marianne BOZONNET, Martin MEIER: 100 Jahre Bakelit. Das Material für 1000 Anwendungen, in: Angewandte Chemie, 120, 2008, S. / pp. 3368-3374.

CROSSCURRENTS PRESS (Hg. / Ed.): The first man in space. The record of Yuri Gagarins historic first venture into cosmic space. A collection of translations from Soviet press reports. New York 1961.

[o.A. / n.author]: DER BANDVENTILATOR, in: Technik für Alle. Monatshefte für Technik und Industrie, 24, 1933/1934, S. / p. 118.

Andrea DINOTO: Art Plastic. Designed for Living. New York 1984.

John Harry DUBOIS: Plastics. Chicago 1943.

Hartmut ESSLINGER: Genial einfach. Die frühen Design-Jahre von Apple. Stuttgart 2014.

Chup FRIEMERT: Hörformen, in: Radioapparate aus der Sammlung Winkler, Ausst.Kat. / Exhibit.cat., Brandenburgische Kunstsammlungen. Cottbus 1996, S. / pp. 13-18.

Philippe GARNER: Sixties Design. Köln / Cologne 2003.

GELSENKIRCHENER BAROCK, Ausst.Kat. / Exhibit.cat., Städtisches Museum Gelsenkirchen. Heidelberg 1991.

Valérie GUILLAUME (Hg. / Ed.): Écrits sur Starck. Publikation anlässlich der Ausstellung „Philippe Starck", Centre Georges Pompidou. Paris 2003.

Sonja GÜNTHER: Die fünfziger Jahre. Innenarchitektur und Wohndesign. Stuttgart 1994.

Thomas HAUFFE: Design. Ein Schnellkurs. Köln / Cologne 2008.

Thiemo HEEG, Dieter RAMS: „Braun hat Apple angeregt – ein Kompliment". Dieter Rams im Gespräch / Designer Dieter Rams in a conversation published, Frankfurter Allgemeine, 27.5.2010, http://www.faz.net/aktuell/technik-motor/computer-internet/designer-dieter-rams-im-gespraech-braun-hat-apple-angeregt-ein-kompliment-1981324.html [17.11.2015].

Michael HENGSTENBERG: Historische Ghettoblaster. Einmal Ohrenschmaus zum Mitnehmen, Spiegel Online, 28.04.2009, http://www.spiegel.de/einestages/historische-ghettoblaster-a-948273.html [25.09.2015].

Hans Jürgen HERINGER: Linguistik nach Saussure. Eine Einführung. Tübingen 2013.

Wolfgang HORNIK: Industriemöbel, in: Sammlerjournal, Juli 2011, S. / pp. 50-66.

Florian HUFNAGL (Hg. / Ed.): Einblicke – Ausblicke: für ein Museum von Morgen. Die Neue Sammlung, Staatliches Museum für angewandte Kunst. München und Stuttgart / Munich and Stuttgart 1996.

Radcliff JOE: Weltron Set 1973 Drive, in: Billboard, 23.12.1972, S. / p. 32 und / and 46.

David JOHNSON, Betty JOHNSON: Guide to Old Radios. Pointers, Pictures and Prices. Radnor 1989.

J. Stewart JOHNSON: American Modern 1925-1940. Design for a New Age. Washington 2000.

Andrea JONISCHKIES: Lauscher am Kühlergrill. Historisches Radiodesign, Spiegel Online, 29.8.2008, http://www.spiegel.de/einestages/historisches-radiodesign-a-947866.html [25.10.2015].

Walter Maria KERSTING: Bilderbuch für Kaufleute. Weimar 1928.

Walter Maria KERSTING: Die Lebendige Form. Serienmodell und Massenfabrikation. Berlin 1932.

Walter Maria KERSTING: Kleine Küche im Großen Haus, in: Die Neue Linie, 9, 1933, S. / pp. 2-3.

Walter Maria KERSTING: KERSTING MARKEN. Essen 1955.

Walter Maria KERSTING: TECHNISCHE GESTALTUNG. [o.O. / n.place] 1955.

Walter Maria KERSTING: ÜBER DEN FORMGEBER, in: Bauen und Wohnen, 10, 1955, S. / pp. 103-144.

Hans Jürgen KOCH, Hermann GLASER: Ganz Ohr. Eine Kulturgeschichte des Radios in Deutschland. Köln, Weimar, Wien / Cologne, Weimar, Vienna 2005.

Hans Ulrich KÖLSCH: „Das Radio brüllte bakeliten". Eine kleine Geschichte des Radiogehäuses, in: Kultur und Technik, 14, Nr. 1, 1990, S. / pp. 2-11.

Hans Ulrich KÖLSCH: Bakelit und Design. Formgebung und frühe Interpreten, in: Ulrich Loeber (Hg. / Ed.): Bakelit. Ein Werkstoff mit Zukunft, Begleitpublikation zur gleichnamigen Ausstellung. Leipzig 1993, S. / pp. 81-93.

Wolfgang KÖNIG: Der Volksempfänger und die Radioindustrie, in: Vierteljahreshefte für Sozial- und Wirtschaftsgeschichte, 90, 2003, S. / pp. 269-289.

Reyer KRAS: Volksempfänger. Walter Maria Kersting, in: Volker Albus, Reyer Kras, Jonathan M. Woodham (Hg. / Eds.): Design! Das 20. Jahrhundert. München [u.a.] / Munich [et.al.] 2000, S. / pp. 56-57.

Friedrich Emil KRAUSS: Haus im Erzgebirge. Schwarzenberg 1933.

Pierre LALLEMAND: Le papier mâché. Sarreguemines 1999.

Günter LATTERMANN: Bauhaus ohne Kunststoffe? Kunststoffe ohne Bauhaus?, in: form+zweck, 20, 2003, S. / pp. 110-127.

Günter LATTERMANN: Der Volksempfänger, in: Kai Buchholz, Klaus Wolbert (Hg. / Eds.): Im Designerpark. Leben in künstlichen Welten. Darmstadt 2004, S. / p. 940.

Günter LATTERMANN: Resopal – weit mehr als Laminat, in: Romana Schneider, Ingeborg Flagge (Hg. / Eds.): Original Resopal. Berlin 2006, S. / pp. 10-19 und / and 193-194.

Friedrich LENGER: Metropolen der Moderne. Eine europäische Geschichte seit 1850. München / Munich 2013.

Niklas LUHMANN: Die Gesellschaft der Gesellschaft. Frankfurt a. M. 1998.

Beate MANSKE, Gudrun SCHOLZ: Täglich in der Hand. Industrieformen von Wilhelm Wagenfeld aus sechs Jahrzehnten. Worpswede 1987.

Walter MEHDORN: Kunstharzpreßstoffe und andere Kunststoffe. Berlin 1939.

Jens MÜLLER, René SPITZ (Hg. / Eds.): HfG Ulm. Kurze Geschichte der Hochschule für Gestaltung. Anmerkungen zum Verhältnis von Design und Politik. Düsseldorf 2014.

Barbara MUNDT: Produkt-Design 1900-1990. Eine Einführung. Berlin 1991.

Marilyn NEUHART: The Story of Eames Furniture, Bd. 1. / Vol. 1, Berlin 2010.

Rudolf NITSCHE, Wilhelm ESCH: Untersuchungen an Eß- und Trinkgeschirren aus Kunstharz-Preßstoffen, in: Kunststoff-Technik, 10, 1940, S. / pp. 57-62.

Jocelyn de NOBLET (Hg. / Ed.): Industrial Design. Reflection of a Century, Ausst.Kat. / Exhibit.cat., Grand Palais. Paris 1993.

Heinz G. PFAENDER, Wiltrud BAUM, Hermann SCHÄFER: Walter Maria Kersting. Architekt, Formgestalter, Ingenieur, Grafiker. Aus seinem Nachlaß. Darmstadt 1974.

Chapman PINCHER: Space Age is here. Soviet satellite circling world in 95 minutes, in: The London Daily Express, 05.10.1957, S. / p. 1.

Bernd POLSTER: Braun. Fifty Years of Design and Innovation. Stuttgart und / and London 2009.

Wolfgang SCHEPERS: Pappe, Plastik und Produkte. Design und Wohnen in einer bewegten Zeit, in: '68 Design und Alltagskultur zwischen Konsum und Konflikt, Ausst.Kat. / Exhibit.cat., Kunstmuseum Düsseldorf. Köln / Cologne 1998, S. / pp. 20-39.

Carmen SCHLIEBE: Legendäres Radio-Design, in: Radioapparate aus der Sammlung Winkler, Ausst.Kat. / Exhibit.cat., Brandenburgische Kunstsammlungen. Cottbus 1996, S. / pp. 19-25.

Carmen SCHLIEBE: Legendäres Radio-Design, in: Luzie Bratner und Gabriele Lueg (Hg. / Eds.): Der 4-eckige Blick. Design und Kunst im Dialog. Highlights einer amerikanischen Privatsammlung, Ausst.Kat. / Exhibit.cat., Museum für Angewandte Kunst Köln. Köln / Cologne 2004.

Eva Maria Josefa SCHMID: Unsere Wohnung. Einrichten und Gestalten. Gütersloh 1960.

Johann N. SCHMIDT: Wolken-Kratzer. Ästhetik & Konstruktion. Köln / Cologne 1991.

Wolfgang SCHMITTEL: design, concept, realisation. Zürich / Zurich 1975.

Eva VON SECKENDORFF: Die Hochschule für Gestaltung in Ulm. Die Gründung 1949-1953 und Ära Max Bill 1953-1957, Diss. Marburg 1985.

Gert SELLE: Design im Alltag. Vom Thonetstuhl zum Mikrochip. Frankfurt a. M. 2007.

Gert SELLE: Geschichte des Design in Deutschland. Frankfurt a. M. [u.a. / et.al.] 2007.

[o.A. / n.author]: SOUND WIRD DESIGN. Bang & Olufsen, in: Novum, Nr. 9, 2011, S. / pp. 20-21.

Penny SPARKE: 100 ans de design. Paris 2002.

Philippe STARCK: Explications. Publikation anlässlich der Ausstellung „Philippe Starck", Centre Georges Pompidou. Paris 2003.

F. J. STÜTNER (Konzept / Concept): Musik wohnlich verpackt. 25 Jahre Rosita Tonmöbel. Paderborn-Schloß Neuhaus 1979.

Sean TOPHAM: Where is my space age? The rise and fall of futuristic design. London, München, New York 2003.

Renate ULMER: Zwischen Art Déco und Stromlinien-Design. Der Durchbruch des Kunststoffs in den zwanziger und dreißiger Jahren, in: Florian Hufnagl (Hg. / Ed.): Plastics + Design, Ausst.Kat. / Exhibit.cat., Die Neue Sammlung, Staatliches Museum für angewandte Kunst. München und Stuttgart / Munich and Stuttgart 1997, S. / pp. 26-35.

Robert VENTURI: Complexity and Contradiction in Architecture. The Museum of Modern Art Papers on Architecture, Bd. 1. / Vol. 1, New York 1966.

Gaston VERMOSEN: Bois Durci. Un plastique naturel / A natural plastic, 1855-1927. Bonheiden [o.J. / n.year 2008].

Hans WICHMANN: Design kontra Art Déco. 1927-1932, Jahrfünft der Wende. München / Munich 1993.

Hans WICHMANN: Mut zum Aufbruch. Erwin Braun, 1921-1992. München und New York / Munich and New York 1998.

Gareth WILLIAMS: Radios, in: Malcolm Baker und Brenda Richardson (Hg. / Ed.): A Grand Design. The Art of the Victoria and Albert Museum, Ausst.Kat. / Exhibit.cat., Victoria and Albert Museum. London 1997, S. / pp. 373-377.

Richard ZIERL: Hundert Jahre illustrierte Radiogeschichte. Geschichte, Entwicklung und Technik. Baden-Baden 2011.

Abbildungsverzeichnis / Picture Credits

Sponsoren / Sponsoring Partners

Medienpartner / Media Partners

ALESSI

Impressum

Colophon

Bestandskatalog des Museums für Angewandte Kunst Köln, Band XXI / Museum of Applied Arts Cologne Collection Catalogues, Vol. XXI

Diese Publikation erscheint anlässlich der Ausstellung „RADIO Zeit. Röhrengeräte, Design-Ikonen, Internetradio" Museum für Angewandte Kunst Köln, 19. Januar – 05. Juni 2016

This publication is published to accompany the exhibition "RADIO Days. Tube Radios, Design Classics, Internet Radio" Museum of Applied Arts Cologne, January 19 – June 5, 2016

Katalog / Catalogue

Herausgeber / Editors
Romana Breuer, Petra Hesse

Redaktion / Editorial staff
Romana Breuer

Gestaltung / Design
Christof Breidenich

Projektmanagement / Project Management, Kerber Verlag
Katrin Meder

Autoren / Authors
Andreas Baumerich, Isabel Brass (IB), Romana Breuer (RB), Petra Hesse, Elina Knorpp, Günter Lattermann, Theresa Nisters

Lektorat / Copyediting
Isabel Brass, Romana Breuer, Susanne Dickel, Petra Hesse, Tobias Wüstenbecker

Übersetzungen / Translations
Susanne Dickel

Die Deutsche Nationalbibliothek verzeichnet diese Publikation in der Deutschen Nationalbibliografie; detaillierte bibliografische Daten sind im Internet über http://www.dnb.de abrufbar.

The Deutsche Nationalbibliothek lists this publication in the Deutsche Nationalbibliografie; detailed bibliographic data are available on the Internet at http://www.dnb.de.

Gesamtherstellung und Vertrieb / Printed and published by
Kerber Verlag, Bielefeld
Windelsbleicher Str. 166–170
33659 Bielefeld
Germany
Tel. +49 (0) 5 21/9 50 08-10
Fax +49 (0) 5 21/9 50 08-88
info@kerberverlag.com

Kerber, US Distribution
D.A.P., Distributed Art Publishers, Inc.
155 Sixth Avenue, 2nd Floor
New York, NY 10013
Tel. +1 (212) 627-1999
Fax +1 (212) 627-9484

Kerber-Publikationen werden weltweit in führenden Buchhandlungen und Museumsshops angeboten (Vertrieb in Europa, Asien, Nord- und Südamerika.

Kerber publications are available in selected bookstores and museum shops worldwide (distributed in Europe, Asia, South and North America).

ISBN 978-3-7356-0175-9
www.kerberverlag.com

Printed in Germany

Ausstellung / Exhibition

Gesamtleitung / Overall direction
Petra Hesse

Kuratorin / Curator
Romana Breuer

Wissenschaftliche Assistenz / Research Assistent
Isabel Brass

Ausstellungsorganisation / Exhibition management
Isabel Brass, Romana Breuer, Petra Hesse, Werner Nett

Grafische Gestaltung / Graphic design
Großgestalten, Köln

Texte / Texts
Isabel Brass, Romana Breuer

Lektorat / Copyediting
Isabel Brass, Romana Breuer, Petra Hesse, Tobias Wüstenbecker

Übersetzungen / Translations
Susanne Dickel

Restaurierung / Conservation
Agnes Eckert, Werner Nett

Leihverkehr / Exhibitis on loan
Dorothée Augel

Ausstellungsaufbau / Exhibition installation
Josef Dreckmann, Krings & Buchal, Werner Nett, Jürgen Plötz

Technik / Technicians
Mike Effelsberg, Frank Schunk, TN Bosch Robert Höschler & Team

Museumspädagogik / Museum Educational Service
Andrea Imig

Kommunikation / Communication
Christine Drabe

Social Media und / and MAKK Design blog
Isabel Brass, Christine Drabe, Kirsten Reinhardt

Veranstaltungsmanagement / Event management
Monika Pfeil

Verwaltung / Administration
Arno Monnig, Diana Richmann

Dokumentation / Documentation
Dorothée Augel

Sekretariat / Secretariat
Hildegard Marquardt

Ein Museum der